The
Stargazing
Year

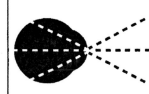

This Large Print Book carries the
Seal of Approval of N.A.V.H.

The Stargazing Year

A Backyard Astronomer's Journey Through the Seasons of the Night Sky

Charles Laird Calia

Thorndike Press • Waterville, Maine

Copyright © 2005 by Charles Laird Calia

Additional copyright information on page 362.

Published in 2005 by arrangement with Jeremy Tarcher, a member of Penguin Group (USA) Inc.

Thorndike Press® Large Print Nonfiction.

The tree indicium is a trademark of Thorndike Press.

The text of this Large Print edition is unabridged.
Other aspects of the book may vary from the original edition.

Set in 16 pt. Plantin by Christina S. Huff.

Printed in the United States on permanent paper.

Library of Congress Cataloging-in-Publication Data

Calia, Charles Laird.
    The stargazing year : a backyard astronomer's journey through the seasons of the night sky / by Charles Laird Calia. — Large print ed.
        p. cm. — (Thorndike Press large print nonfiction)
    Originally published: New York : Jeremy P. Tarcher/Penguin, c2005.
    ISBN 0-7862-8110-3 (lg. print : hc : alk. paper)
    1. Stars — Popular works. 2. Constellations — Popular works. I. Title. II. Thorndike Press large print nonfiction series.
QB801.6.C35 2005b
  523.8—dc22   2005020968

For my Father

As the Founder/CEO of NAVH, the only national health agency solely devoted to those who, although not totally blind, have an eye disease which could lead to serious visual impairment, I am pleased to recognize Thorndike Press* as one of the leading publishers in the large print field.

Founded in 1954 in San Francisco to prepare large print textbooks for partially seeing children, NAVH became the pioneer and standard setting agency in the preparation of large type.

Today, those publishers who meet our standards carry the prestigious "Seal of Approval" indicating high quality large print. We are delighted that Thorndike Press is one of the publishers whose titles meet these standards. We are also pleased to recognize the significant contribution Thorndike Press is making in this important and growing field.

Lorraine H. Marchi, L.H.D.
Founder/CEO
NAVH

* Thorndike Press encompasses the following imprints: Thorndike, Wheeler, Walker and Large Print Press.

We are all in the gutter, but some of us are looking at the stars.

— Oscar Wilde

God has arranged everything in the universe in consideration of everything else.

— Meister Eckhart

There is no peace like the peace of a starlit evening.

In the cool stillness of my backyard observatory the quiet has a signature. A mechanical hum, faint and guttural, like a swarm of unhappy bees, drifts from the telescope as the motor drive locates a new star. But the hum is not all that I hear. Dried leaves rustle, yet there is no wind. The woods are alive with creatures of the night — mouse and deer, raccoon and skunk, fox and owl. I have seen them all. Hours spent in the dark have made my eyes sensitive to every movement, and my nose, weaned on clear, country air, pulls in the slightest hint of odor.

Above me are stars. Thousands of them. The evening is clear and it is late. All my neighbors are asleep and all the lights are extinguished, caging the dark like an animal. A few hours earlier the sky had a faint white glow near the southern horizon, light pollution, spillover from local houses, outdoor lights, cities, but now even the light pollution surrenders. All that I am left with is the night and the terrible weight of the universe with its mysteries. But the real mystery, I'm slowly discovering, is inside of me.

I first looked through a telescope when I was eight years old. Astronomy sustained me through a turbulent adolescence that I shared with many of my generation. In America we are all children of change, technological and social, and I wasn't any different from most of my peers except for what I did at night. I looked at the stars. A telescope was as common to me as a pair of shoes, and just as functional. Something necessary. I lived this way all through high school, when it wasn't always easy to justify my passion, and into college.

Then suddenly I gave it up.

For most of my adult life I simply ignored the call to look skyward; instead I looked only straight ahead. The gravitational pull was a natural one, and the trajectory clean — career, family, children, a mortgage. The orbit of small lives, I've found, is as dramatic as the motion of great celestial bodies, and the same laws apply. Motion tends to remain constant until halted in midcourse.

What stopped me was something that I

saw on television. I couldn't explain the events of that September morning but I knew one thing: I felt solace when I looked at the stars.

The stars have changed little since the dawn of human history. But then, we are looking at the stars of our past, not the present. The light that it took for these objects to reach us varies, a snapshot in time that has already unfolded except that we haven't witnessed it yet. This constant, the idea that nothing ever changes, is an illusion, of course, like the illusion of safety or the feeling that a life measured and carefully planned turns out as expected. It doesn't always. The stars do change, in small measures, as we change in larger ones.

In the autumn of 2001, I beat a circuitous route back to the love of my youth. It began with a tour of the night sky with my two daughters, both young, with the freshness that comes only with new eyes. My oldest daughter got the ball rolling.

"See how clear it is, Daddy."

We were pulling into the garage, our car full of heavy shopping bags from evening errands, and as usual I drove in fast without bothering to look up. Why look up? But for some reason I slowed the car, cracked the window. Fresh air overtook me and I felt it. Absolute clarity.

I spent that evening lying on the grass with my two daughters, pointing out the constellations, whose outlines and names, oddly enough, I still remembered. Like everyone that autumn, I was feeling confused and angry, afraid for my children and my nation and disoriented by the stories that I was hearing in my own community, from friends and neighbors who worked in Manhattan. But that night, gazing at clumps of broken Milky Way, a thought washed over me, one that I have yet to surrender. We all belong to the sky.

October is usually the clearest month in Connecticut, where I live, and the night darkens with the fainter stars of autumn. The heat of the summer gives way to cool Canadian air that slips down on my state like a cold splash of mountain water on the skin, cleansing everything and making senses tingle. Leaves turn, clouds fatten into white balls, and at night the constellations appear above me, sentinels of change.

It was Cygnus, the great Northern Cross, that welcomed me back that night with my children. High overhead, at the head and foot of the cross, are two stars — Deneb, the bright supergiant, and Albireo, one of the finest binaries, or double stars, in the heavens, and they framed the entry of my return to astronomy. To the south of Deneb, the brightest star in the cross and the nineteenth-brightest star in the night sky, gave me Altair in the constellation of Aquila. A poor man's version of Cygnus, Aquila, the eagle, spreads out its wings reluctantly, and if not for the majesty of Altair herself, which is brighter even than Deneb, the constellation might hardly be noticed at all except for one thing. Altair is the pointer for Vega.

Vega, the brightest star of the summer and autumn skies, lords over its recessional in the west. It is also part of a great and yawning **V** of stars, a triumvirate, with Deneb and Vega at the base and Altair manning the point, alone, a cop on a stakeout.

This is the famous summer triangle of my youth.

As a teenager growing up in the early '70s, I used this familiar marker as a road map to navigate the sky at night. From Vega I could march, as a general taking strategic towns and bridges would, right to the small parallelogram of stars in Lyra where two objects lay in wait: the Ring Nebula and the famous double-double.

The Ring Nebula remains a testament to

the violence of the cosmos. Violence has shaped much of what surrounds us on earth, and the heavens are no different. A blasted shell, the nebula's small, transparent ring haunts the surrounding star field with the memory of what was. Long ago, the central star within the Ring Nebula began to shed its luminescence. Gasping for breath, the star burned more hydrogen, the loss of which made the exterior swell up, heaving off layers of gas like so many clothes on a Florida vacation, until all the layers and all the hydrogen were gone. What remains is frightening: a circle, not unlike the ghostly puff from a dying cigar, where eons ago a star must have shined brightly.

But not all is gloomy in Lyra. Just north of the Ring Nebula a cosmic dance takes place nightly every summer and fall in telescopes all over the world. This is the double-double, a multiple star system that tricks the eye not once or twice but three times. Visually the star looks like any other moderately faint star. But with a pair of binoculars one notices a celestial twin. Increase the magnification and the punch line of the joke is revealed: two star systems, each one a small pair, pose for the observer like diamonds on a jeweler's black velvet cushion. These are binaries, double stars orbiting around one another in a vast and intricate waltz that is best seen with binoculars or a telescope. But looking up that October night I realized this: I had neither.

Twenty-five years earlier, in order to help pay for a semester of college, I sold the telescope of my youth. It was a green Discoverer, the finest that Sears offered, a 60-millimeter catalog refractor that my father had bought for me when I was twelve. The gift was an expensive one for him and no doubt foolish — a tool-and-die maker, my father saw little utility for such an extravagant instrument beyond the call of familial love.

"Don't you already have a telescope?" he asked me.

I did. A cheap one.

"And now you want a second?"

All eyebrows. It was difficult for my father to trust the need for two of anything. He was raised in Kentucky during the Great Depression: a single suit and tie, one pair of school shoes and one for chores, a set of overalls and farm clothes. Clearly a telescope was a frill, unnecessary. He could understand, all too clearly, the desire for hand tools, even a baseball glove. But a telescope?

"I need it, Dad. It's more powerful."

That great temptation. Power.

"With it I can see more."

My father took his time to consider this in that measured way that children of the Depression often do, weighing the costs and benefits. But I was impatient. I knew if I was going to own another telescope anytime soon I would have to close the deal, and fast, which meant going to my mother.

It was easy to find her. She was usually in her "office," as she called her bedroom. She spent most of the day there, propped up in her bed, writing on a lap desk that looked like it came from Marat's tub. My mother rarely left her bed, not because she couldn't but because, I thought, the universe inside her mind was so compelling, more interesting than the mundane world outside her door. This intrigued many of her friends, including her doctor, a large, block-shaped man who met with my mother twice a month for ten years. He enjoyed their conversations so much that they often veered into the mystical.

"He's a Leo," she said. "Very eccentric man. Lots of problems."

My mother was an astrologer. Next to her nightstand she kept a stack of notebooks, essays and articles that she had written, and ledgers too, on friends and family, with each

16

of our birthdays plotted and drawn up into charts. On the charts were symbols that I recognized, ancient shorthand for the planets and their movement through the zodiac, except that her zodiac was largely constructed from imagination. The planets are seldom where astrologers place them in the sky, and we argued about this constantly.

"Jupiter isn't in Aries, Mom — it's in Cancer. Look up."

"I have."

She tapped a blue book called *The American Ephemeris.* In it were the astrological charts and tables that she used to calculate the destiny of her children, which she purported to know, or perhaps it was only the instinct that all parents have about their kids, an eerie prescience.

"You want something, don't you?"

It didn't take a soothsayer to know that.

"I need a better telescope," I said.

"Really? You know —"

Here comes the company line, I thought. Not money or lack of it, but astrology. For centuries astronomers have battled with astrologers, arguing the veracity of science by pointing out that astrology gets everything wrong. But the war has continued, and nowhere was this more evident than in my own home.

"— astrologers were the first astronomers. Chaldeans."

"Science, Mom. Not superstition."

She smiled faintly. "The stars will always be a part of your life," she said, "one way or the other. It's in your chart, so I might as well embrace it. Let me talk to your father."

The new telescope arrived a few weeks later and it soon occupied the most prominent place in our house, which was the dining room. While most folks ate in quarters decorated with family pictures or art, my family dined under the gaze of the largest moon map I could buy, almost six feet in diameter, tacked strategically between two windows.

My parents didn't care much for convention and almost seemed to relish the quirkiness of the map. Besides, the moon was in. Only months earlier, I had sat up with my family and watched Neil Armstrong on television, in black and white, a bad daguerreotype descending from the LEM. Now the moon was the center of our home.

I'm not sure exactly where my father, a practical man, stood on the subject of the moon. Maybe he found comfort in the fact that our country was exploring new territory or maybe he was alarmed by all the whimsical metaphors attached to this old world. Wishing on the moon. Pie in the sky. Moon swoon. On weekends he would try to corner me to help with his basement project, a complete overhaul of our downstairs into the classic '70s version of the *Brady Bunch* family room — wood-paneled walls and matching bar, stools, and shag carpeting, a real party spot despite the fact that we seldom entertained.

"I could use a hand with some Sheetrock," he said, "if you can spare one."

"Planetarium, Dad. Sorry."

In lieu of changing oil or learning carpentry, useful pursuits in life, I spent every weekend playing sports or at the local museum. The museum was my afternoon haunt. It offered a planetarium show, for one thing, which was the same week after week for months on end, but I didn't care. There were stars, and I practiced identifying the constellations as the dome went cold and the Pennsylvania Dutch landscape, where I grew up, silhouetted itself on a fake, cardboard horizon. It was home, that planetarium.

At night, after the planetarium show, I would pull out the telescope and observe until my parents ordered me in. Rarely was it before eleven, and it was later on weekends. By high school, I was staying up until all hours, early dawn sometimes, and dragging myself into first-period Latin where I took the opportunity to catch up on my rest.

My teachers knew about the astronomy, of course, and the observations that I was making, charting weather patterns on Jupiter, and a few gave me a break. Most saw it as a harmless albeit eccentric hobby, but some felt it was the pastime of a dreamer, probably confirming my father's worst nightmare, that astrology and astronomy were linked, indeed were one discipline, and that I would imagine my life away while romancing the stars.

Even my friends saw that. Weekends were the worst, and the parties that ensued required the obligatory late evening. I often used astronomy as an excuse, like protecting a secret from a jealous girlfriend. I simply couldn't be gone on a clear night. The ribbing would come.

"What do you do out there? Spy on chicks?"

Somehow, when I actually told people, it came out sounding more ridiculous than even I thought possible. I was learning the night sky, I said, which elicited howls of laughter.

"Sorority girls, you mean. They live behind your house."

The local college.

"Like you're really looking at the stars all night with those hot bods."

Whether anyone ever did believe me or they all just tolerated my hobby because it was amusing, I'll never know. As it stood, I was allowed to flourish in my neighborhood, like a quirky uncle, an eccentric who fortunately could pitch.

"Okay, Mr. Cosmos. Show us the Apollo flag, then."

A common misconception. Non-astronomers invest even the smallest instrument with remarkable gifts. They can find astronauts, view the surfaces of tumbling asteroids, peer into reaches that the largest telescopes in the world have trouble seeing — all for a few hundred bucks.

It simply isn't true.

"I can't," I said, "but I can show you their landing site."

Sure, I was exaggerating. I could, in 1970, locate the landing area of the first few Apollo missions within a fifty-mile radius, sort of like finding Manhattan from space without actually seeing a building on Times Square.

In astronomy, size matters. All amateur stargazers, and most professionals as well, are forever conducting mental trade-offs between the equipment that they own and the equipment that they want. As a boy, I was firmly in this camp: owning what I no longer desired and desiring what I couldn't own. A seductive twist on the Cistercian line of wanting what I had: I simply wanted more, and I skirted around the problems by making friends with larger telescopes, volunteering at observatories, going to conventions; in short, doing whatever I could to be around bigger instruments.

Now, years later, while looking at the night sky with my children and trying to explain the Ring Nebula instead of merely pointing a telescope at it and letting them see for themselves, I found myself stewing, as a jilted girlfriend still angry a decade after the fact would. The man who bought my childhood telescope was a lumberjack who lived in northern Minnesota. He owned a cabin in the darkest part of the state, just south of Ontario, near the Boundary Waters, and he wanted a telescope — mine — but I raised the price three times in the span of five minutes, hoping that he would lose interest.

He didn't.

With a thick wad of bills, the man purchased my past: all the hours that I had spent with that eyepiece, and more. Perhaps reeling from my despair, I decided to give him everything that I owned pertaining to matters astronomical — books, old and valuable star atlases, filters, measuring devices, extra eyepieces, the whole kit and caboodle — and with one sweeping financial transaction I found myself in a new land, the land of memory. I was clean, healed of my obsession.

I believed this for nearly a quarter century.

Winter

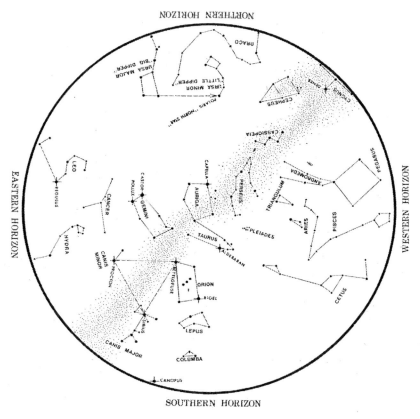

THE NIGHT SKY IN JANUARY

January

The idea came as many ideas do, unexpectedly, like a bright meteor across the night sky. A few months had passed since that evening spent outside with my children. I used most of my time working, trying to forget the clarity of that fine October night, but I wasn't very successful at it. For one thing, I had a pair of new binoculars and was busy, on clear evenings, wheeling them through constellations like Captain Ahab on the high seas, except that I wasn't in pursuit of some demonic whale but rather open clusters, bright variable stars, lunar craters. Soon the idea that I had took precedence.

It began with a chair and a cold, clear night. The chair was inexpensive, a standard gray Wal-Mart job, and it allowed anyone stupid enough to sit on it in January the experience of frostbite. It was eight degrees outside and the mercury was falling fast. That afternoon there had been snow, almost three inches added to the pile we already had on the ground, and ice crystals were twinkling off the snowpack in the dark. Lights from the neighboring house took these crystals and twirled them around as if they were a mad kaleidoscope, hinting at the spectrum — red, orange, yellow — the Roy of Roy G. Biv high school physics.

The hours that I spent outside, acclimating myself to the heavens, went fast. Like most obsessed people, I told myself that I was fine, that I could handle it, but the cold penetrating my gloves and coat that evening seemed to suggest otherwise. I was crazy. Worse, I didn't care. I was sitting outside in the jaws of a deep freeze, losing track of the minutes, as if this were normal. It wasn't. I heard the window open upstairs, and the voice of my wife, checking to see that I hadn't frozen or, more likely, slipped down an icy ravine, only confirmed this.

"Are you still out there?" she asked.

My response, a muffled affirmative voiced through Bedouin wrappings of wool and Thermolite, alarmed her.

"Do you need a blanket? Hot chocolate?"

She looked down from the warmth of the house. Ice fell and crashed from the window.

"I don't understand the allure," she said, closing it.

Who could?

January is the coldest month of the year in New England. But what nature takes away with the promise of summer warmth it more than makes up for by offering its greatest visual prize to the backyard astronomer: its clarity. The atmosphere is a transparent dome of moving air. High, thin clouds can dim the stars, as can the murky nights of spring, but in winter the sky often appears scrubbed and the stars twinkle with clean abandon — not always good for views through a telescope, though stirring to the naked eye. Now a front was moving in and the air was crisp. The snow storm of that afternoon had been carried off the coast of Nantucket, and the winter constellations were peeking through the last of the clouds like Christmas lights overhead.

Orion, the great hunter, is easily the most recognizable. The stars of the hunter form an hourglass with a curved shield. On his belt — three distinct stars in a small row — there is a knife such as any good hunter would own, and in his hand he wields a mighty club. There is danger in the night sky tonight, as the bull Taurus looms above Orion. The hunter stands his ground, like a reluctant matador who fights not out of some time-honored tradition but because he must. Taurus threatens everything around him. His horns are distinctive, a great chevron, like the flight of autumnal birds, that stretches northeast from Aldebaran. As the brightest star in Taurus, Aldebaran gleams with the red eye of the bull — incensed, angry, primal — and tells Orion one thing: I am coming for you.

Orion, thankfully, is well armed. He has protection, the long shield of the Spartans, and with his club held high above his head, he is eager to strike. Once Taurus is stunned, Orion can then reach for his belt and finish the job with dagger or sword.

The sword of Orion, a blur of light to the naked eye, is really a stellar nursery. Through binoculars or a telescope, this blur becomes one of the most dramatic sights in the heavens. This is the Great Nebula. A filmy patch of ionized hydrogen, the nebula is home to a small grouping of four stars, the Trapezium, four newborns in a universe aching to grow up, develop planets, pull in the surrounding gas of the sword, and grow old. I suspect it's the dream of all living things, sentient or otherwise; the hope of fulfilling what nature has designed for them.

The gas in the Orion nebula glows with a primeval eeriness. It's focused at two points, one within the Great Nebula that fans out and fills most eyepieces to the brim with ribbons of light and dark, and the other around the western part of the belt itself. The star is called Zeta Orionis, after the Greek naming system for star identification invented in 1603 by Johannes Bayer for his book *Uranometria*. Bayer used the Greek alphabet to categorize stars in a constellation according to their brightness, or magnitude, with the most brilliant given the letter Alpha on down a descending order to Omega.

The sixth brightest star in Orion, then, is Zeta Orionis, though the Arabs, eminent astronomers in their own right, had another name for it. Alnitak. The girdle. A girdle it is, for Zeta Orionis restrains around its luminosity one of the most famous dark nebulas of the sky — the Horsehead. This nebula is nearly impossible to see except with a specially attuned filter or on photographic film. Its dark imprint of a small head, profiled like that of a child's hobbyhorse, appears as black water, an eddy in the white foam of rapids.

The Horsehead Nebula is unique. The gas is actually absorbing the light around the stellar nursery, filtering out a view that has become classic in astronomical photography and is probably recognized by more non-astronomers than any other sight out-side our solar system.

It's while looking at the area around the Horsehead that I get my idea. Sneaky, covert, the idea slithers in the night air.

What if I could take a picture of this?

Ideas are like bubbles, floating around for anyone to grasp. They are also like germs. They begin singly, one thought or a variation, and soon they start multiplying. My idea already had another one behind it, neatly stacked like shark teeth. Photography was little more than an excuse to open up an entire cavalcade of even more expensive thoughts. To properly photograph the Horsehead, I decided, I would need something more than binoculars. I would need a telescope.

Time can condense in a second within our memories. The novelist William Faulkner, sitting on the back porch of his comfortable home, summed up best the shakiness of what we consider to be historical. "The past isn't dead," he wrote, "it isn't even past."

Whether the cold overcame me or the idea had simply been knocking about in my head since I was a teenager, I can't say, but the connection between past and present was as sure as a link in a chain. Where would I put this telescope, I mused? And the answer came in the night.

An observatory. I would build an observatory.

If every amateur astronomer hungers for a more powerful telescope, then what they hunger for next is a place to put it. Fields, patios, and gardens are fine for small telescopes and can easily accommodate a shelter, even a common one.

Backyard observatories come in all shapes and sizes. They are often simple, constructed of wood and unfinished on the inside, without Sheetrock, unlike the popular conception of the domed observatory, its roof shining in the desert. Some amateur creations, though, are elaborate works, with computers running charge-coupled devices, or CCD cameras, photoelectric photometers, and bleeps of WWV blaring on the shortwave. Backyard astronomers are engineers, cabdrivers, writers. And as such they carry with them the variety of gifts that God gave them, or talents that they learned from their fathers, the army, or trade school. Some astronomers had the worst of all possible worlds, avoiding all three. They spent time at the planetarium instead of helping with the basement and learning real skills like carpentry, a fact that tickles my father thirty years later.

"So you finally want to build something."

Like most men of his generation raised on a farm, my father could fix a roof, do some plumbing, renovate rooms. After the war, he painted houses in Florida before settling on his real love, which was designing boxes for special transport — an inspiration, no doubt, for my father continued to build and design all throughout my childhood. He made a mobile kitchen for family camping trips, homemade tools and scientific instruments for me, shelves for my mother's astrology books. My parents divorced when I was in college, and my mother moved to Seattle, as far away from Pennsylvania as she could get, while my father hightailed it to a golf-course condominium where he continues to tinker — re-gripping his golf clubs while he talks to his son about building.

"My advice? Talk to Jim."

Jim is my brother-in-law. A replica of my father, he can build anything.

"I want to do it myself, Dad. My own two hands."

"Not everyone's a carpenter, you know."

This was an understatement. I bruise my thumbs nailing up pictures.

"I can learn."

"Well . . ." A slow, Kentucky pause. "Building's hard. Are you sure?"

I nod. But his grin says everything.

Pie in the sky.

Most amateurs don't reach this point. They rent, own small yards, have wives who complain or refuse a monstrosity like a shed that spoils the view; or else they are more sensible. Telescopes go back into the house after they're used, like my own Sears refractor; into living rooms and basements all across this fair nation. Telescopes are multiplying in America and they are multiplying fast.

Observatories are not far behind. Thirty years ago they were rare sightings. No longer. Domes and sheds are popping up at an alarming rate, matching an expanding hobby. The reasons for building one are varied, and not hard to articulate, at least to another backyard astronomer. Simply put, a telescope ready to go is one that often gets used. An observatory helps with this, obviating the lugging out of equipment and the constant searching for something more permanent. A building keeps out bugs, dew, and wind; it shields the observer from stray light, and keeps the telescope's alignment in check. But most important, a building protects the telescope. Good reasons to build an observatory, perhaps, though it is not quite as easy as keeping the telescope in the basement.

But storage in the basement offers problems of its own, especially in winter and particularly in cold northern climes like that of Connecticut. For a telescope to work properly, its optics must reach an ambient temperature with the outdoors — a cooldown period. Failure to do so means the image will suffer, boiling away in the eyepiece as the glass throws off radiant heat, a process that can consume an evening.

This sounds easy enough. Just stick it outside and wait. But menaces lurk. Dew in the summer, ice and frost in the winter. Covers help. Low-level heaters around the optics work even better. So why not wait? Depending on the telescope design, waiting is an option. Refractors, the classic lens in front of a long tube, tend to cool down faster than most scopes, but they are often unwieldy and long. They are also limited in their apertures, or size; roughly, how fat a telescope is corresponds to its ability to gather light. In the world of astronomy, unlike in Weight Watchers, airplanes, and carnival rides, fat is good. An unnecessarily long and slender telescope suffers when collecting light.

The chubby Newtonian reflector, on the other hand, is an instrument designed for speed. It can be wheeled out quickly, set up, and calibrated in no time, but its size is often a hindrance. It's a large telescope whose mirrors need careful alignment, more than any other telescope's, and the hauling and lifting of this instrument can rattle its delicate calibrations.

This is hardly a problem with the most popular telescope in America. The sturdy Schmidt-Cassegrain is such a ubiquitous design that it can now be purchased at the shopping mall, in camera stores, and at discount chains. While even Schmidt-Cassegrains need careful alignment, they tend to hold their alignment for longer periods, thanks to a glass corrector plate that prevents the secondary mirror from shifting. But glass works both ways. It traps in heat as well as allows in starlight, a process of thermal cooldown that can take up to an hour in cold weather.

A home observatory eliminates all this work. The telescope is cooled down and ready for action within minutes of opening the roof, a point that I make to my father when he asks me what I'm thinking.

I can tell he isn't sure about this. An observatory? His eyes narrow, pulling in the lines of his brow, deep furrows that I remember from when I was a kid. He asks me about my writing. Books.

"You're building those," he said. "Besides, you don't own a telescope."

Not yet, I thought.

Professional observatory selection committees, when determining the efficacy of a project, usually don't proceed this way, by thoughtlessly jumping in. First they find the necessary financial backing, securing donors and grants, and then they move to the design of the instrument, finding and testing a site, all before construction — a process that, in the case of the Palomar Observatory, took 19 years. Selecting the telescope helps to determine the type of observatory, not vice versa, the way that I was doing it. To build an observatory, I first had to consider what was going inside it. The telescope.

It began with the Internet. After a few days of surfing, I was bug-eyed and bleary, but I had catalogs, PDF files, specifications, and user testimonials piled high on my desk, with still more coming from the printer. Since my childhood, telescope making has become a big business. You can always gauge the health of amateur astronomy from the number of telescope makers. It was true in the nineteenth century and it holds true now.

As a boy, I saw the growth of two titans: Meade and Celestron. These commercial powerhouses are now two of the largest manufacturers of telescopes in the world. Not then. They were youngsters, startups almost, battling the established makers: Unitron, Criterion, Cave. And then there was Questar.

I had one wish in 1974. It wasn't the wish of most sixteen-year-olds, a used car; my wish, one that I dreamt of daily, was a new Questar telescope. I could have just as easily asked for the used car. The prices were about the same in those days and my chances of actually getting one was equally remote. Good cars were expensive. Questars were even worse.

What made this design so special was a carefully figured meniscus-lens-and-mirror combination that produced refractor-quality images, extremely sharp and contrasted, but without the bulk of a refractor — thanks to a folded optical system. Light was first corrected through the glass meniscus, then bounced from a primary mirror back to the lens, where a small reflector, fixed on the back of the meniscus, was lying in wait. One last volley sent light careening into the eyepiece. It was a pinball machine for photons.

The execution only takes an instant, of course, exactly the dream of Dmitri Maksutov when, in 1941, he designed and built his first meniscus-lens–based telescope in Russia. Elegant and compact, Maksutovs looked like telescopes in miniature, but their prices were far from miniature. They were aimed at rich men.

For a sixteen-year-old amateur astronomer, there were only two options for a Maksutov telescope: Quantum and Questar. Both offered precision optics in a package that could fit on a desk. The scopes were tiny, sure, but with the finest optics in the world. Not surprisingly, Questar was still around thirty years later, and they were still expensive — many thousands of dollars. I glanced at my depleted bank account. Not today.

Astronomical products have changed much through the years, but no change has been more dramatic than that of the number of telescopes now available. Telescope makers have sprouted like mushrooms since my college days, as my generation returned to the love of their childhood hobbies for capitalist inspiration.

Telescopes are also flooding in from China. And these telescopes are not just cheaper than any were when I was a boy; they're more sophisticated as well. I felt like a man returning from a space flight to discover that people no longer drove cars. The effects were revolutionary, mind-boggling. But then, I always have been living in the past.

A confession: when I was a boy and observing with my simple Sears refractor, I was already a dinosaur. Astronomy had been changing even then, first at the professional level, with computers and CCD cameras, and then among the amateur forces when we saw what the pros were using. Amateurs wanted a piece of the rock. Now we have it, in scale.

Computers have revolutionized everything, so it's no surprise that they have also revolutionized backyard astronomy. No longer must an enthusiast pull out his charts to find objects. Now it's point-and-click. The computer inside the mount does all the work, moving the instrument to the desired location in seconds. Unlike many hobbies, amateur astronomy requires a great deal of knowledge. In order to learn how to use a telescope, you first had to learn the night sky — but knowledge of the night sky requires using a telescope — a catch-22.

No longer. Computers have taken the guesswork out of astronomy. A person new to the sport merely has to set up his telescope, point it at the North Star and a second, well-known comparison object, and go. Instant knowledge. Many of these new telescopes will even tell you what you are looking at with a brief running commentary, like a docent standing beside you at the museum. My mouth watered at the catalogs.

But the more I thought about a computer doing the work for me, the less I liked it. Part of the challenge in finding faint celestial objects is knowing where and when to look. The knowledge is passed along, initiate-style, with tips from experienced observers, and from star atlases. I wasn't interested in losing that. I wished to find my own objects, sweat over them if need be, if only to say I'd done it.

We humans have a peculiar way of latching onto the past and never letting go, but the future is a good thing. Change sweeps in like a cold front, signaling that life carries on, renews itself. Dinosaurs are relegated to the ashbin of history. Luddites remain so until they change. My peace with technology would come later, but on that cold January evening, hatching the plans for the observatory, I had one desire: to reclaim my past, and find meaning in the whimsy. It began with a catalog from California. They had new, inexpensive telescopes, identical to the Questars except at a fraction of the price. I ordered one.

And then I did it. Maybe it was the observatory I was thinking about, or perhaps I was only imitating steps familiar to me, I'm not sure. But suddenly I made a second decision. It was the telescope for the observatory. I wouldn't use the Maksutov as my main instrument, I decided; I would buy a second, even larger telescope — the instrument of my youth.

I would buy a refractor.

We purchase our past. The motto is an old one, known by antique dealers. If this is true, then my decision to buy a refractor would mean buying not only my past but the early past of observational astronomy.

A refractor is the classic image of a telescope. With its long tube, its objective lens in front, and a mere peephole for an eyepiece, this is the instrument of Galileo. There are two mysteries surrounding the refracting telescope. One is who actually invented it, and the other is why it took so long to invent at all.

Spectacles with lenses that refracted light

had been around since the fourteenth century, but it wasn't until 1608 that a Dutch optician named Hans Lippershey demonstrated the first working telescope. Lippershey then did what any decent entrepreneur would do: he showed his telescopes to people who could buy them in quantity — in this case, the States General in The Hague. The idea was a good one, so good that he applied for a thirty-year patent.

He didn't get it.

The telescope spread quickly around Europe. It was knocked off, reproduced, improved upon, leapfrogging from Amsterdam to Paris to the court of Henry IV. A letter from Jacques Badovere, a former student of Galileo's, arrived in the hands of the Italian scientist with incredible news. There was an instrument capable of bringing distant objects into closer view. Was he interested?

Galileo took a toy that could magnify only a few times and made it into an observing platform from which to explore the moon, sun, and planets. He constantly redesigned the telescope and may have made more improvements to the refractor than all his predecessors combined. He elongated the tube, created more powerful eyepieces, and most important, he built a mount to steady the instrument. With it, he observed the moons of Jupiter — the ones that got him in so much trouble with the Church.

The refracting telescope continued to improve after Galileo. The width of the objective lens, or aperture, slowly widened, and the focal length, or overall distance from the main lens to the eyepiece, also increased. Astronomical instruments in the seventeenth century were long affairs, aerial telescopes, often reaching a hundred feet in length or longer. They were cumbersome, necessitating the use of ropes and pulleys, and an observer was in danger of being crushed by his own instrument.

There was also a question of quality. Galileo used two lenses of poor optical value, one for the objective lens and a second at the eyepiece. This caused problems with false color — colors not natural to the objects viewed — but more critically, it also caused aberrations within the system. Objects rarely came to a clean focus, and those that did still showed much distortion.

A telescope maker named Fraunhofer solved this problem by introducing a second piece of glass in the objective lens. This shortened the length of the telescope and cleaned up many of the aberrations, and telescopes got wider. The wider the telescope, the better, and in the eighteenth and nineteenth centuries, aperture width began to accelerate. Soon there were refractors more than 24 inches in diameter, an amazing size given the precision it took to cut and collimate, or align, the lens, and discoveries began to pile up.

But refractors quickly fell out of favor. It was the cost and limitations of their objective lenses that sealed their doom. In the seventeenth century, Sir Isaac Newton modernized optics with the invention of a new telescope — the Newtonian reflector. This model soon found its way into nearly every large observatory in the world. A deep, bowl-shaped mirror captured light from a short tube and bounced it off another mirror, or secondary, into the eyepiece, not bending it like a refractor would but reflecting it. The idea was revolutionary. Mirrors were first made from a metal called speculum, later ground from plate glass, and they could also be manufactured easier and more cheaply than lenses. They could be built much larger, with the ability to catch more light.

This thirst for size has a name. Amateurs call it aperture fever. Spouses call it grounds for divorce. Soon, stargazers were grinding their own speculum. One of these amateurs, a German musician by the name of William Herschel, was one of the best grinders of speculum mirrors in all of England. He was also one of the finest astronomers in the world. In 1781, he discovered the planet Uranus from his backyard and was given the title of Royal Astronomer, which came with something that Herschel greatly desired — a stipend. This offered him the chance to dedicate himself full-time to astronomy, an obsession that had by then drawn in his sister, Caroline, and he would go about using his assets wisely.

Herschel didn't build an observatory per se; his entire house was one. In it he had several telescopes, including his favorite, a 20-foot reflector that he designed himself, an instrument that was rumored to be the best in the world. With it, he discovered the satellites of Uranus, found comets, and charted the heavens with the zeal and passion of a man possessed. He was.

Two centuries ago, an amateur observer was largely up against himself and his optics. There were few professional astronomers back then, and the history of astronomy, until the twentieth century, is really the history of amateur astronomy; hobbyists with great curiosity. Now much of that has changed. Professional astronomers are in control of most of the science going on today, although an amateur still has something to contribute. After all, there are only so many telescopes and so much time. Amateurs have made inroads in the study of variable stars, novas, and eclipsing binaries — inroads that have proven, even today, to be immensely valuable to scientists. So too are the observations of weather patterns on Jupiter, the discovery of comets, and the study and tracking of asteroids — all within the domain of the advanced amateur astronomer in his own backyard.

Throughout the world, on any given night, thousands of people direct their telescopes at the heavens. Many are casual observers, but an equal number are serious students of astronomy. They read, keeping up on their particular area of interest as any professional would, with this exception: these people are not paid. Perhaps this is for the best. Love and money rarely mix, and it is an unusual person who would work for free but people do. Some wish to contribute something to science; others are genuinely curious; some are frustrated wannabes. But most of the amateurs who spend countless hours at the low end of an eyepiece do so, I would suspect, for one simple reason: Love. They love the feel of the night and the peacefulness under the stars, and it is this love that signals to us what we all need to know.

We are wondrously alive.

My first telescope in twenty-five years arrives a few weeks later, like many new telescopes seem to, in the worst weather imaginable.

The ice cold of early January has turned into the deep freeze of late January. It is, some folks say, one of the coldest months ever. Old-timers jog their memories and conjure obscure facts and figures, mostly relics, fossils of memory, and decide that this winter is merely normal. Like the old days. New England was always cold, they say, and the warmth of the past few decades has been an aberration. If so, I long for the aberration to continue.

For the first few days that I own my telescope, the sky is frozen shut. Clouds don't move, or if they do, it's only to bring in more clouds, dump snow. A great, cold nebula has seized the entire Eastern Seaboard for almost a week now, and I am relegated to sitting in my garage and waiting. The wait, however, is a useful one. I lubricate the mount and bearings and practice setting up the telescope. The mountings are equatorial, balanced to the celestial equator, a cockeyed **T**, and I'm familiar with its workings, using the angle of my latitude, 41 degrees north, and a powered motor drive to simulate the motion of the earth. Because the earth is always moving, a telescope has to do the same, and yet it must stand firmly planted in the ground.

Most observatories employ a steel pier, fixed to a slab of sunken concrete, for their mounting. This was my plan exactly, but now I must wait. The ground is frozen and won't thaw out until spring, precious months that I use to reacquaint myself with the heavens.

Soon the clouds lift, revealing the double star cluster in Perseus. With the naked eye the view isn't very exciting: a dim blur, or rather, two patches, stacked like a double oven. The eye alone can't resolve these objects, but binoculars can. A telescope goes a step further. The hazy patches explode into dual vessels of spackled light that fan out, tumble into one another. A wide-angle eyepiece puts both of these clusters into the same field of view, and the blackness between stars is dark. Spilt ink.

My viewing is interrupted by a sound in the woods. I live in the middle of a forest — ten acres of dense trees with a hole punched overhead. A clearing. The forest is sanctuary for a variety of creatures. There were rumors last summer of a bear roaming these woods — a rumor that few can verify, so people add other tales, linking them as if to give the bear story credence: coyotes and even a cougar.

The cougar sighting was taken seriously. A State of Connecticut naturalist actually came out to investigate more than one claim that a cougar was seen wandering across the nearest country road, not far from town. The naturalist took notes, not unlike a good amateur astronomer would, comparing his data, establishing evidence. A cougar hadn't been seen this far south, this close to Manhattan, in more than fifty years, he said, and he was skeptical. He spoke with several neighbors before concluding that it must have been a bobcat — rare in itself — but this didn't quiet the neighbors any. They locked up their cats and kept dogs on a tighter leash, and always there was a worry about children.

Me, I wasn't worried about a cougar or even a hungry bear. What had me concerned was another kind of creature that I saw all summer long.

Skunks.

Already I've had encounters at night.

More than once, a passing skunk has stared me down, his tail twitching, as I attempted to become invisible, fearing the worst: burning my clothes, scrubbing my skin with lye, and brewing down-home recipes from my father's Appalachian childhood. But the skunk simply turned away, deeming me unworthy of his time. This evening, though, I was prepared.

I flashed my red penlight into the woods, and I saw movement in the snow. A mole or rodent of some kind. The light has its limitations. I painted its outer bulb with red nail polish. Red preserves the night vision of stargazers, and everything around the garage and exterior now has the tint.

"You don't think it looks like a bordello?" my wife asks.

It actually looks like an observatory. Observatories all across the world employ low-level lighting, often red, to help astronomers, infrared cameras, and CCD detectors adapt to darkness, though my red light isn't very reliable for identifying creatures.

Animals are stargazers too. I've noticed that many of them become more active on those crisp and cold nights when the stars dance and you get the feeling that you can see forever. Sometimes you can. If you look hard enough, you can see almost to the beginning of time — human time at least. You can easily look back two million years just by glancing skyward.

As I look past the double cluster in Perseus, I notice, for perhaps the first time ever, the Andromeda galaxy, and I can see it with my naked eye. This is a real feat under modern, light-polluted skies. A small stain, the outline of another galaxy beyond ours, that is two million light-years away, seems pretty innocuous. But in a photograph, the great spiral of the galaxy comes to life. Tight spiral arms and a condensed core makes this galaxy a companion to our own, with its dust lanes and smaller, satellite galaxies.

In a telescope, little of this is seen, though the galaxy fills most of the eyepiece. Andromeda is bright. The core of the galaxy resembles a cotton ball, and around the ball are faint elliptical streamers, the galaxy's arms, that resemble closely our own neighborhood. Earth is situated on the edge of the Milky Way galaxy, not too far out, not too close to the center — a Goldilocks position that others might envy. Our block of nine planets and the sun is a humble one, hardly the dominant role envisioned by the early Church. We are a star among billions, a minion among minions. The mystery of our origins is as varied as the mysteries in our lives. We could have ended up anywhere, and yet here we are, on this living rock that we call Earth in a galaxy scarcely distinguishable from any other beyond the fact that we call it home.

Home is not always an easy place.

As a kid growing up in a cramped, industrial Pennsylvania town, I perceived my home to be a hindrance to astronomy. First, the skies were murky, and the row house that I was raised in, with its postage-stamp backyard and maple tree, hardly allowed much sky. Then there was the isolation. I was born into a working-class neighborhood that betrayed the arrogance of my youth. What could these barbers, tool-and-die makers, milkmen, and industrial welders possibly teach me about astronomy, I thought?

Everything.

The welder across the street showed me my first solar eclipse through his welding glasses. He read about it in the newspaper and we set up on the sidewalk, passing his dark goggles back and forth. Soon there was a crowd, and I discovered that our local car mechanic built imaginary rockets in his basement, with flashing lights and working toggle switches. He wasn't alone. Another neighbor with a machine shop inspired me to construct a camera adaptor, which actually worked. There was also Judy, our next-door neighbor, who coolly watched my interest in astronomy grow for years before surprising me with her stack of old *Sky & Telescope* magazines from the '50s. I pored over this pile for months on end, and it gave me a solid foundation for my hobby.

Judy, I discovered, was an amateur astronomer. She owned a small telescope, hardly much larger than a spyglass, which she had used as a kid to search for comets. She never went to college for astronomy, as far as I knew, though her sky knowledge was astounding and in every respect professional.

The fact that I had been living next door to an amateur astronomer for years on end and didn't know it rattled me when I first found out. But it didn't rattle my mother. For her, the metaphors of sky and earth were linked, and we had only to open our eyes to discover it.

"Everything is connected," she said. "You're shocked?"

I was. "What are the odds?"

"There are no odds, that's the point."

I never bought it, her cosmic determinism. A pair of dice rolled so many times came up with predictable numbers, I argued, so the fact that my next-door neighbor was interested in astronomy only confirmed the best scientific hypothesis that I could muster at that tender age.

"There must be a lot of folks interested in astronomy," I said.

My mother just smiled. Her obsession with the stars, parallel with my own, continued long into her life, and when she died, unaware of my return to astronomy that summer of my stargazing year, the box that was delivered from her estate to my house had all the earmarks of cheap fiction — a plot device.

It was a telescope.

And yet, despite her astrology or perhaps because of it, she was an inspiration for my own astronomy, as was my father. He designed and built measuring devices for me as I got older, filar micrometers, and he taught me that anything that could be imagined was transferable to paper and then reality — the trade of the tool-and-die maker and the artist.

He also taught me to make the most of my surroundings. We lived near a college, with its planetarium and small observatory within walking distance, and something else. A library. I spent hours in the stacks, sifting through unintelligible scientific papers, British journals, astronomical log-books. In short, I did whatever I could to stay close to the thing that I loved the most. Naturally the assumption in my own home and around the neighborhood was an obvious one. I would go to college, become an astronomer.

Nothing could have been further from the truth.

An admission. After high school, I honestly intended to study astronomy at college, but something happened. I couldn't handle it. I blamed leaving school on the advanced mathematics involved, the endless classes in physics that made my head spin, though in reality these were just excuses. I was diverted from astronomy not by a woman or hard work, but by a book. *The Stranger* by Albert Camus.

I knew Camus' work. My mother was a great and voracious reader with an extensive library in French existentialist literature, though I found this ironic. She didn't believe much in free will, other than as a cursory exercise, I think, which was perhaps why she was so frustrated as a writer and why I was steering so close to it. I refused to read Camus or much of my mother's literature until I was a senior in high school, when the fear of becoming an adult was just beginning to settle in. But the book came highly recommended by a writing teacher whose words I took to heart.

"You'll like Camus," he said.

Camus spelled doom for my astronomy career. I just didn't know it yet.

When I returned home from college the next summer, the decision had already been made. I wanted to be a writer. I could hear the disappointment in people's voices.

"I thought you wanted to go into astronomy."

Expectations are investments, and no investment is greater than the love of parents for their children, especially the love of a practical father.

"Won't you need to earn a living?"

"I'll be a writing astronomer," I announced. "Both."

My father's eyebrows went up. More pie in the sky.

In retrospect, what I told my father was exactly the way that I felt. I had always envisioned myself as a writing astronomer, or an astronomer who happens to write, not unlike Carl Sagan or Patrick Moore. But my father was wise. He knew that a man couldn't have two temptresses and that I would be lured away by the craziest whimsy in the world — the life of the imagination.

The questions in cosmology and fiction are similar. They deal with origins, the thread and arc of a life: whether it's a character or a galaxy, it doesn't matter. The motions are the same. Life, a muddling middle age, death. The uncertainties that cloud the eyes and judgment of men also cloud galaxies. They too have their crises. Some collide, merge, fall victim to gravity. They lose their shells of hydrogen, as fundamental as a man losing his mind to Alzheimer's. Galaxies roll up, dissipate, exhaust themselves. Men do this as well. Our hydrogen is our youth, and the promise that tomorrow will exceed our expectations. It doesn't always.

The reality of leaving astronomy at such an early age took an unforeseen toll on me. I quit looking skyward, and the whimsy that a writer banks on soon evaporated. After the first novel came the permanent second novel. Words began to drag, and then even the passion that I'd relied upon vanished. It was time for renewal. It was time to look at the stars.

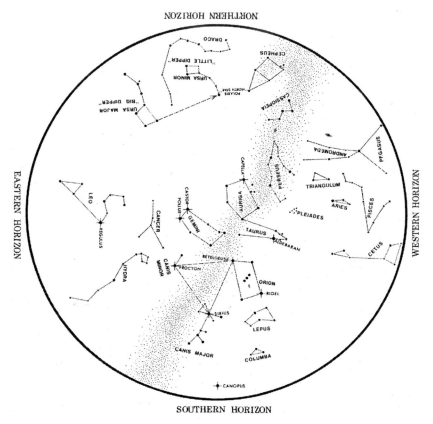

THE NIGHT SKY IN FEBRUARY

60

February

Geometry rules the winter night. First the angles of snowflakes, intricate and original configurations that all children know, and then the stars. From my south to the zenith overhead, I can see not just one triangle, like in late summer and autumn, but two — six stars neatly stacked in editions of three, base to base, with room to spare. This is the great winter Hexagon, consisting of Aldebaran, the bull's-eye; Capella, the goat star, at the zenith; Castor to the east; Rigel, at the knee of Orion; Sirius, the nose of the great dog; and Procyon, the nippy mutt.

For the past week now it has been clear almost every night, a miracle, and I have been spending it outside with the new telescope. I also have a mobile observatory, consisting of a cart with wheels, that I push, as a shopper in a department store would, past the sale items to the lure of the expensive stuff.

Since Christmas, the products have been pouring into the house as I try every piece of astronomical equipment that I can afford, attempting to make up for lost years. Years cost money. My bank account was suffering, digits dwindling, and the smolder of credit cards was already visible at the other end of the mailbox — a rising plume of obsession. Wives notice smoldering credit cards.

"UPS was here. That's twice this week already."

I nod, shuffle through the house. When I'm lucky, my wife isn't home and the gear can be assimilated quietly. I suspect I'm not alone here. Like men driving their expectant wives to the delivery room, we amateur astronomers can have a problem with limits. A constant flow of new toys doesn't help any.

"How much is this eyepiece?" asks my wife.

It's the size of a coffee mug. She inspects it as I quiver.

"Looks pricey."

It is. More than a French dinner for the family, but I can't admit that.

"Not bad," I say.

My code is broken. Eyepieces, my wife soon learns, have gone straight up in price along with cars, college educations, and the bills at French restaurants.

She nods. "I just hope you use this stuff."

I do. Outside, in the dark, I roll out a

second table. I need two. On this one I keep my charts, a newly minted observing log, multiple eyepieces, a case to protect them, filters, a scramble of pencils, dew zappers, two batteries, a laptop computer loaded with the latest astronomical software — the list goes on. What is remarkable is that I'm not alone. During any given evening in this country and around the world, there are thousands of others shadowing my efforts. They set up mobile observatories or ready fixed ones, one ear to the weather station and an eye on the sky to confirm it.

This routine has been going on for hundreds of years, certainly since the advent of telescopic astronomy. Amateur scientists creep outdoors, after tucking in their children, kissing wives, and walking the family pets, to a sky made glorious with stars, and no month exhibits a better sky than does February.

Dogs dominate. Every hunter has one, and Orion, the greatest of hunters, is no exception. He has two by his side: Canis Major, which actually resembles a dog, with its four legs and pointed snout, and Canis Minor, which does not. To say that it is a stretch to imagine Canis Minor as a small dog is to state the obvious. The stars of Canis Minor appear as an afterthought, it seems, a kind of ad hoc assembly designed to give the constellations the appearance of balance. Or perhaps Canis Minor exists only as a pointer for Sirius.

As the brightest fixed star in the sky, Sirius commands Connecticut's southern horizon every winter. Twinkling Christmas-tree blue and white, the luminescence is almost enough to cast a faint shadow on the ground. Called the Great Star of Isis by the Egyptians, Sirius was also the Nile Star, the Heavenly Wolf to the Chinese, the Dog Star to the Babylonians. Every culture has a name for this star, and most of them offered respect. Sirius demands respect.

It also has a star to announce it. Procyon is known colloquially as the Little Dog Star — a kind of sidekick for Sirius. But this star is no slouch. As the eighth-brightest fixed object in the sky, Procyon anchors the eastern edge of the Hexagon and leads the way towards Castor, one of the famous stars of the Twins.

Gemini appears in the sky as two **L**-shaped lines set back to back. These are supposed to be the shoulders and feet of the children, who are apparently hugging each other with outstretched arms. But they resemble more the bodies of conjoined twins. The stars of the head, Pollux, the brightest, and Castor, a nice double star that is nearly as brilliant, form a mirror identity, a doppelgänger, suggesting to me that the ancients saw Gemini as a delightful metaphor.

They really were twins.

Gemini is home to several wonderful deep-sky objects, including the open cluster M35. The French astronomer Charles Messier cataloged this in 1784 as part of a growing list of 109 objects that he felt resembled the blurry haze of comets. These were rocks in the sky for Messier — sheer nuisance, or, worse, treacherous ground for the comet hunter. Rocks crush boats. They also stop astronomers cold as the hunters scramble to their desks to chart what had always been there — distant light.

Messier was a comet hunter, one of the best, a man who notched more than a dozen comet discoveries himself. He swept the sky nightly for evidence of bright objects just beginning their orbits around the sun. Their gases ignited, these comets would glow in the visual spectrum and Messier would nab them, a lone marshal arresting fugitives, except that these comets weren't the famous fugitives that he had hoped for. Long forgotten for his comet work, Messier is now known not for the objects he discovered but for his list, beloved by amateurs across the globe. The Messier list targets bright objects in the northern sky, and M35 is one of these.

Seen through binoculars or a small telescope, the cluster resembles a sugar cube dissolving in water. The edges break away first, revealing stars that flee the center. Finding M35 is easy. Simply locate Castor, the second-brightest star in Gemini, then sweep toward the foot in a straight line and move a thumb-length north. You can't miss it.

If you continue north from Castor to the tip of the apex, you reach Capella. This star dominates one of my favorite constellations, Auriga, whose poetic name and angular, obelisk-shaped grouping of bright stars betrays its mythological significance — as the Charioteer. Within the boundaries of Auriga are three more prominent Messier objects, open star clusters, and these too are easy to find, especially given the dimensions of the winter Hexagon.

But the geometry of the sky suggests to me a larger theme this night, one that I notice at the legs of my telescope. What if I built an observatory to match?

My wife, I've discovered, isn't keen on geometry. She also isn't keen on the idea of an astronomical shack in our backyard.

"You want to build *what?*"

"An observatory with six sides, maybe three."

She looks at me and grins.

"Fix the stairs first, then we'll talk."

But I press my point.

"You've never built anything in your life," she says.

My silence worries her, I'm sure. Wives have radar. They can sense the inner wheels, the plotting and calculating in male brains. At least, my wife is used to mine. Only months earlier she talked me out of buying an antique letterpress.

"Besides, you don't own any tools."

"Oh, I can get those."

Giggles turn to incredulity. Soon I feel it. The great thumb.

"You're not going to build an eyesore in our yard. Forget about triangles, hexagons, pentagons, or anything else weird," she says. "I know you. Halfway into this project you'll lose interest and rush to get it done. Then it'll look like —"

Her mouth twists the words.

"— the laundry room. The floor still buckles."

She was right. My first attempt, as a new homeowner all those years ago, was to redo the laundry room. I tore out old cabinets and ripped up the floor with glee and a demented zeal. Men like projects. We enjoy the thought of tearing something out and replacing it, even if we don't always know what we're doing. The unknown is part of it. Mettle is tested. Plans are concocted and put into action. Modern existence has isolated many suburban men from the rigors of safari life. No longer do we go on the hunt, keep the fire, spear a dangerous beast.

Now we safari at Home Depot.

But doing the laundry room made me realize this: I have no skills. I would like to blame going to college, but I can't. The truth is, I know a dozen other educated men who can paint a house, do carpentry, fix transmissions, build banjos. All I can do is read.

"How are you planning on doing this?" she asks.

I shrug. "Books. I've seen them at the hardware store."

I'm back to reading again.

But my wife knows this. The fact that I even went to the hardware store means something. An idea is in my head: the observatory. Whether shed, hut, or octagon, the idea is tugging at me, and I relent to the gravitational pull of most of my schemes unless reasoned with first.

She tries. She catalogs my skills, lovingly and with great marital tact, which takes about ten seconds. I can mow grass, operate a leaf blower, change a few screens and maybe a windshield-wiper blade, but only after much swearing and consternation. I can also locate stars, but she doesn't mention that. The message, of course, is clear. These are not skills.

On television I would be the first guy voted off the island. Ten thousand years ago, my tribe might have considered me a menace to the food supply and banished me, bait for the local saber-toothed tiger. In other words, expendable. I have gleaned, in my years on earth, few practical abilities. Worse, I seem to be regressing. Skills that I learned in jobs long ago have vanished from memory. Disuse does that. Rust does that as well. And no amount of convincing my wife that I could build the observatory would help. I needed to convince myself first.

It is perhaps not surprising that astronomical observatories have always been around. Nearly every culture on every continent, with the possible exception of cultureless Antarctica, has spent an inordinate amount of time looking skyward. They all did this for a variety of reasons, many of which still resonate with us today. People look up to measure the time, to notice the phases of the moon, or just to stand in awe of the mystery.

The first observatories were probably the mouths of caves. But as humankind progressed, a need arose, in which the religious practices of the day were complicit, to unify these effects as a scientific discipline. Observatories were created then to observe the sky and report back the findings to a waiting populace.

The oldest observatory in the world might not be one at all. The pyramids of Giza were constructed around 2650 B.C. by slave labor, as heavy stones were lugged from the quarries at Luxor up the Nile. For years, archaeologists have noted that there was a connection between the pyramids and the sky, particularly to the constellation of Orion, who was seen by the ancient Egyptians as the god Osiris, the god of the Pharaohs. And one explanation even suggests that the physical layout of Giza itself resembles Orion, with the three main pyramids as the belt and two on each side as the incomplete arm and leg of the hunter. But is this an observatory?

Our modern conception of an observatory is really an old one. It hinges on some degree of technology, a scientific method to observe and make verifiable calculations; a method that even the ancients practiced.

Take Stonehenge.

Archaeologists debate over the dating of Stonehenge, built probably between 3000 B.C. and 1900 B.C., but they don't argue over how long it took. This was a transgenerational engineering project, one conducted over many decades for the express purpose of calculating astronomical events. Stonehenge was a megalithic observatory then capable of computing, with frightening precision, specific lunar and solar incidents, no doubt including eclipses. It was also an accurate calendar useful in agriculture, religious and political rituals, and historic celebrations.

In 1740, William Stukeley noticed that the main axis of the large stones, or megaliths, aligned itself to the midsummer equinox, which suggested a larger purpose for Stonehenge than the accepted notion that it was a temple. Stonehenge was also an observatory, he said.

The American astronomer Gerald Hawkins confirmed this. He noticed that megaliths acted like a gigantic slide rule, using a series of holes, stones, and arches as a kind of advanced geometric model of the heavens, thus allowing the user to predict the rising and setting of the moon and sun with unprecedented accuracy. This gave the architects of Stonehenge a fresh, almost intimate relationship with the sky, one achieved through a world that is particularly modern. The world of mathematics.

The use of mathematics, particularly geometry, abounds in many of the ancient megalithic sites that are found across England, Ireland, Wales, and Scotland. To the eyes of an amateur astronomer, these are backyard observatories of the most passionate kind, involving long man-hours of construction, planning, and observing, all for the reason of embracing the stars.

Stonehenge wasn't the only early observatory. Around the world are the scattered remains of many such places. In modern day Mexico there is the great temple of Caracol, a site dedicated to the observation of Venus. In China there is Guo Shoujing, an observatory built in the eleventh century and that helped determine the length of the calendar year. But few observatories, either those still standing or those long destroyed, have stirred the imagination more than Samarkand, a facility that was once located in what is now Uzbekistan.

The Samarkand observatory, a building

said to be more than three stories tall and manned by a small army of scientists, was the largest in the world, built by an amateur astronomer who also happened to be a prince. His name was Ulūgh Beg, and he was born into the violence of the age, an unending stream of military campaigns and political infighting on the Asian frontier, a milieu created by the Mongols and Kublai Khan. But the young prince's heart was captured early when he looked skyward and decided to make his mark with the sciences. He did. He built an observatory.

From 1420 to 1437, the Samarkand observatory cataloged more than a thousand stars, the most since Ptolemy, by employing the largest stone sextant ever constructed. With a radius of 120 feet, the Fakhri sextant, as it was called, could measure the position of a star within a few seconds of arc, equal to the diameter of a penny from half a mile away — a remarkable achievement. This was truly a working observatory with fixed instruments, a new concept in observatory design then, that allowed users to calculate the length of the solar year to within fifty seconds of our modern measurements.

Amazingly, the Samarkand observatory did all this work with simple instruments and the naked eye. Telescopes and telescopic observatories were still almost two hundred years in the future when Ulūgh Beg articulated his dream, the dream of many astronomers since: an observatory to unlock the secrets of night.

Of all the secrets of night, no question was perhaps more important to the ancients than this: how much do the stars influence our everyday lives?

My mother had her view. They influence everything.

She was born in 1925 in Chesapeake, Ohio, the daughter of an astrologer, not a stargazer. There were no telescopes in my mother's family, nor were there astronomers, but there was a fervent interest in the heavens. Both my grandfather and mother wore a ring that had stars embedded in it, neat diamonds that sparkled in the light, as stars do, on a background of black onyx. But these were symbols, I was told: nothing so real as to be charted in the actual sky, they were connections to a universe that, ironically, nobody ever looked at.

And yet real stars held powerful sway in the family. I recall a visit to my grandfather's house when I was a young boy, during which I quickly noticed something unusual about his chimney. There was a star in the brick, etched in copper, its five points glimmering with the rays of the rising sun. Which star was it? Sirius? Vega?

My mother shook her head.

"That's the Eastern Star," she said mysteriously.

Even as a kid I knew the Eastern Star was really just Venus, in the morning skies, not a star at all, and I told her that.

"How do you know it's Venus?"

"That's east. Venus," I said. "The morning star."

She nodded. "Trust me. It isn't Venus."

So the Eastern Star remained an enigma to me. But this was the first time I saw a gulf between mysticism and science; between astrology and astronomy. Before then, astrology to me was simply the daylight pursuit of the heavens, only on paper, perhaps no different than the calculations that professional astronomers made in the classroom. An astrologer uses the same language of the backyard stargazer — conjunctions, oppositions, transits — but with implications, I discovered, that were far more perilous.

When my mother moved to Seattle after she and my father divorced, I recall her consulting her charts, not for the best apartments, the most vibrant bookstores, or poetry readings, but to see if the stars were cosmically aligned. Seattle is a tough place to be a real astronomer, with its rain and murky nights, though the heavens of the imagination are far easier to wade through. They're always clear.

My mother never thought about telescopes much when I was growing up. She had no interest in them at all, but she did find a fascination with my books and star charts, whose mathematical tables and computations in degrees and seconds of arc rivaled her own. There was an exception, though: I could look up. Astrological charts are metaphors, not reality. Uranus is never physically in the same constellation as where the astrologers place it, and I reminded her of this. When I became a teenager, our arguments were rarely about grades, friends, or staying out late; they were about the heavens. I evoked the power of science, exactness, and truth, against what I considered a fraudulent incursion.

"If you can't locate the planets," I said, "then what about your charts?"

They weren't very accurate, I suggested, meaning that all bets were off. Predictions, suggestions of planetary convergences, cosmic stereotypes.

A smile would cross her lips.

"The actual position of the planets is immaterial. It's their influence that matters."

And here she revealed herself. For every celestial event, not unlike Newton's law of gravity, there was an action that warranted an equal and sometimes linked reaction. It wasn't enough for the planets to exist; they also had to command mysterious forces.

"You're talking about chunks of rock and gas, you know that."

"I'm talking about ideas," she said.

Back and forth we would go. On a few nights I would pull her out to the telescope, point it skyward, and show her the error of her astrological ways. The constellations always cooperated, even though she didn't recognize them. These were paper constellations for her, rough outlines of something conceptual. For me, the sky was alive, and the marvels that I was showing her were witnesses of the real mystery of creation — why there was so much of it. The universe just kept going, an Energizer bunny on steroids, the sky leaches into everything that we see and even far into what we cannot.

"If you want riddles," I said, "look up."

When did the universe begin? How will it end? Can it end, or will matter on the fringes of the Big Bang just peter out? One night the questions rolled off my tongue, both small and large, unanswerable and mildly interesting like those of a good amateur scientist. What science didn't know was far more enigmatic and strange than what could ever be imagined, and I told her that.

She smiled, nodded. I was confirming her argument, of course.

"Watch out. You're getting mystical with your astronomy."

It took me half a lifetime to understand what she meant.

One of the most famous observatories in America was also founded by an amateur. Percival Lowell, a Harvard-educated mathematician, was born into a wealthy Boston family in 1855. Raised in proper society, Lowell drifted like most young men of means; he traveled the world in search of his destiny. Expectations were high in the Lowell household. There were Lowells in business and education, including a brother, Lawrence, the president of Harvard University, and his sister, Amy, a poet who would go on to receive the Pulitzer prize. The bar had been set high by his siblings. But Percival Lowell was looking even higher. He was looking at the stars.

Lowell had been interested in astronomy since he was a boy, and he took a six-inch telescope with him when he went to Korea and Japan as part of a diplomatic assignment. But he didn't do much observing. On his way home, he stopped in Arizona, and maybe it was there, under the steady, desert skies, that he got his idea to build an observatory. In 1894, construction began above the hills of Flagstaff, with space to accommodate his new instrument, a 24-inch Alvan Clark refractor that Lowell had gotten on special order from the telescope maker himself.

Alvan Clark and his sons were master opticians, and their name and equipment are still revered throughout the astronomical community. It was Clark who polished the largest refracting lens in the world, a massive 40-inch objective near Williams Bay, Wisconsin, and his instruments were said to be the finest available. Refractors were specialists, long telescopes capable of resolving fine detail on planets thanks to their focal length, and this was exactly what Percival was looking for. He was interested in the planets, and especially in one of them.

Mars.

Since the beginning of human history, Mars has inspired a special fascination. Perhaps it's the color, like a drop of blood hanging in the sky, that led to all the forbidding names given by the ancients: the Burning Coal, the Fire Star, the Death Star. Mars was a world in conflict far before Percival Lowell set eyes on it, though he would create the largest stir yet.

It began with an Italian astronomer, Giovanni Schiaparelli, who in 1877 observed a network of thin lines crisscrossing the planet. Schiaparelli dubbed these lines *canali,* or grooves, but the English translation that came to Lowell was something more fanciful. Canals.

Words may shape reality, and they clearly shaped Percival Lowell's imagination. That *canali* meant canals suggested a larger framework. Artificial waterways, like a gigantic Martian public-works project, brought water down from the planet's polar ice cap to where it was needed most. The deserts. Through a small telescope, Mars appears like a crimson Earth. There are fixed landmasses, continents at first blush, that are bordered by dramatic and restless red deserts, sweeping dust, clouds, and morning mists across the thin Martian atmosphere.

The polar caps tend to recede in the summer, a slow melting that sparks the most controversial of Martian events — the greening of Mars. Lowell must have seen this effect, and no doubt it confirmed his faith in the *canali,* as the planet, parched and dry, moves the last remaining water to farms, fields, and oases. Today we know that this is an illusion, an albedo — or brightness effect — caused by shifting sands, but in 1894 this wasn't certain. Mars was assumed by many astronomers to be a living planet, perhaps even a populated one, and there was no more vocal proponent of this theory than Percival Lowell.

It can be a dangerous enthusiasm, the passion of an amateur. Passion can move mountains, build observatories. It also can veer into fancy. Lowell took his story of a living Mars on the road to anyone who would listen. He lectured, wrote books and popular articles, gave interviews to magazines and newspapers; in short, Lowell was the Pied Piper of the canal theory of Mars, namely, that an intelligent race of beings needed water from the poles to quench their thirst. This theory would find its way into the popular culture, helping to launch, or at least validate, a rising genre. The science-fiction novel.

The War of the Worlds, the classic H. G. Wells novel, whether or not it was inspired by Lowell's conjectures, took the world by storm in 1898 and helped to define the argument. We may not be alone. Worse, we may even be in danger. Earth was a planet of water, and if, as Lowell thought possible, there were living Martians, they would be in desperate need of it. What if someone wanted ours?

Fiction does this. It advances ideas by suggestion, like the voice of a good hypnotist, taking our hopes and fears and shaking them, turning out their pockets. Soon Mars fever was everywhere, and small telescopes were being purchased at an amazing rate. Observatories were being built as well, on college campuses across America and in the backyards of the wealthy, as the red planet became a new obsession. What observers saw split the astronomical community in half. Some sided with the canal camp, while others saw nothing even close to canals. This controversy raged until 1965, when *Mariner 4* rewrote history. Mars wasn't a desert oasis at all but just a desert. Its pockmarked surface, pelted by meteors, pushed up by volcanoes, and split by fissures many times the size of the Grand Canyon, resembled the American Southwest without vegetation more than it did a lush, green world.

Mars appeared dead.

But the romance of the Red Planet remains. As a boy during the great opposition of 1971, when Mars was at its closest approach in a decade, I remember looking at the planet with my Sears Discoverer. The recognizable detail was scant. I could pick up the polar ice caps, a bit of the wedge of Syrtis Major, a dark, **V**-shaped marking near the center of the planet, and some light areas including the surrounding desert, especially a fingerprint-shaped smudge called the Hellas, but little else. There were no canals for me. And yet, every night I went outside and spent an hour or more at the eyepiece, hoping for what I can't exactly say. The canals had long been relegated to the ash heap of history, an optical illusion to the thousands of observers who claimed to see them. The sky plays tricks on the eye as it plays tricks on the heart. But it is the heart that pulls the observer out night after night. Science isn't always romantic, but romance surely has its draw, and without it, all human endeavors, including the thirst to know, wash up on a shore drier than even the Martian desert.

Perhaps Percival Lowell knew this, for he clung to the myth of the canals even when the science of his own day was turning against him. Or maybe he was simply acknowledging the romance within him, the power and curiosity of youth that sustains an astronomer in his backyard, at night, alone in the dusk with only the stars over his shoulder. One thing is clear: Percival Lowell took his dream and made it real. The observatory still stands and remains a testament for dreamers everywhere.

My own dream of building an observatory crystallized with a book. As a teenager I sold my entire library. These books were old, often heavily illustrated tomes with passages that I had underlined in yellow marker, some of which were so familiar to me that I still knew them by heart. Of course, much of the information was now useless, outdated or revised. Science moves quickly, and books are often the first casualty, but the thirst to know guarantees a steady stream, the young replacing the old, though some work becomes classic.

One such classic was a book entitled *Astronomical Telescopes and Observatories for Amateurs.* The author was British astronomer Patrick Moore.

"Once a really powerful telescope has been obtained," Moore wrote, "the problem of a permanent observatory has to be faced. Taking a telescope to pieces at the end of each observing session is a wearisome business, and also a dangerous one; it is only a question of time when something is dropped, with disastrous results."

As an author and backyard stargazer, Patrick Moore walked the walk as well as talked the talk. On his grounds in Selsey, England, Moore had built not one observatory but three, each photographed for the book at different angles, that lovingly represents the three different options for the backyard astronomer.

On the high end were domes, rotating cupolas constructed of metal, canvas, plywood, or prefab plastic. A few of these observatories, built by other amateurs, were attached to raised decks or even to the tops of houses, though I found this curious. Any heat is detrimental to observing, and the warmth from a house, whether from the furnace or the constant expanding and contracting of shingles, make image quality problematic. An observatory must be unheated, and as much as we amateurs would like to observe in our short sleeves in a Vermont January, it simply isn't advisable.

While observatory walls can be constructed of a variety of materials, the floor should be wood. Wood is the simplest and cheapest of building supplies, and it easily surrenders to the elements. An observatory will lock in some heat or cold during the day, but it shouldn't radiate heat throughout the night. Wood also helps when it is used for framing around the pier, a metal pole cemented into the ground to stabilize the instrument. All telescopes need to be mounted properly, but a telescope fixed in an observatory needs the best of all foundations, which is concrete. A sonotube, like those hollowed-out cylinders used in the construction industry, will do the trick, provided that it is positioned below the frost line.

The pier should also be separated from the observatory floor, isolating vibrations. Telescopes vibrate, and even the strongest and most solidly mounted of instruments suffers from some movement. In the eyepiece, every vibration is magnified to a wobbly cuneiform of irritating light, and it is the job of a good observatory designer to focus on pier management.

Moore seems to have taken that job seriously. I see a concrete pad at the foot of another observatory, one of the easiest kinds to construct. This is the run-off shed. It consists of a wooden shed on wheels that can be pushed away from the telescope, thus exposing both observer and telescope to the perils of the night. The design was inexpensive, but I didn't like it. The astronomer was prey to the weather. Bugs could still swoop in; ice and dew will form more easily. Most of all, the observer had to contend with lights. A structure with walls could block some neighboring windows and help with fighting off the evening moisture, I figured, and it was here that I went to the picture of Moore's third observatory. The roll-off roof.

This was a simple shed, like those purchased at home garden stores around the country, but with one big difference. The roof gave way to the stars. For the amateur astronomer, the variety of roof designs are as diverse as people. There are roofs that fold down, lift off, roll, or shift like line dancers, and a good designer can make any of these methods work. It looked so easy. Go to the building-supply store and examine a shed, copy it, then tear off the rafters. What could be so tough about that?

The lot adjacent to the store was full of toolsheds and miniature houses for kids, but no plans. I tried asking for help, flagging down a salesman in an orange vest to assist me, or at least to let me pick his brain for tips.

"You can't build one of these for the price," he said, pointing to the deluxe model. It was a shed that came with windows, a weathervane, and planter boxes. "Even if you have the tools, it's a bargain. Do you have tools?"

I had a hammer, screwdrivers, a hacksaw, and an awl.

The salesman rolled his eyes. These tools were fine for the eighteenth century, not now. He told me that I would need a cordless drill, circular saw, jigsaw, ten-inch miter saw, a T-square, two sawhorses, a caulk gun, caulk, bags of screws, galvanized nails, and a nail gun, and only then would I be getting started.

"You haven't got to the ground. God help you there."

It wasn't flat.

"First you have to bring in soil. Then you'll need to pack it down. A good truck will do. Don't forget your gravel and a level. Get to know it well."

He handed me a flyer for the garden shed.

"My advice? Buy this or hire someone."

I left the store depressed. The project was much more complicated than I'd anticipated. This meant that a hexagonal structure was clearly out. Even a triangle wasn't an option. I would be lucky, I thought, if I could cut straight lines, which meant a box. And the smaller the box, the better. Ten by ten feet. Eight by eight feet. I scratched off the plan in front of Jim, my visiting brother-in-law, who knew the problem before I did.

"Your roof. What are you going to do about that?"

I slumped over. It had to be easily removable.

"Can it fold up?"

"And be watertight?"

"Then I'll roll it off."

A faint smile. "By hand? Pick up a few bags of roofing shingles."

Heavy.

"So it pops off, Jim. Think moonroof with a chain."

"That's some chain. What about gravity?"

Gravity would pull the roof to the ground, he said, with an instructor's voice. At least not without a counterweight and lots of testing.

"Something has to support the roof. You like science. Think about it."

I like fiction as well, and this project was sounding a lot like it.

Most carpenters spend a considerable amount of time on roofs. But few homeowners do. We are worried about our roofs only when they are being built, shingled, or when they leak. Otherwise, we hardly give them a thought. Rain pelts our houses, wind pummels the exteriors, snow heaps in mounds, and yet the roof is a constant, the viral protection software of our house, it's always up and running. Roofs are also meant to stay on houses. Storms, tornadoes, and hurricanes will test any structure, but a homeowner should not. And neither should amateur astronomers. For a carpenter, the idea of constructing a house with a detachable roof was a lot like asking a surgeon to cut out only the healthy organs. Simply unthinkable.

But observatories are built on a single premise. The roof opens up. Some amateurs effect this by using motors and wheels to pull the roof from its moorings, allowing a small area for the telescope to peek out. Others simply roll the roof out onto gigantic supporting beams cemented in the ground. These beams, both the size of the footprint they required and the sheer work involved, frightened me.

"Why can't the roof support itself?"

"It's off, remember?"

"Then I'll stack it. Imagine shuffling a deck of cards."

He just shook his head.

"You'll need to keep the water out first."

The water. He was right. The idea of water seeping into the telescope was the worst scenario imaginable. Certainly the objective lens should be sealed, capped up, but moisture has a way of creeping into everything. I imagined myself observing with a towel in one hand and a blow-dryer in the other.

"Then there's the footings. Concrete."

"Concrete?"

"For the foundation."

My head starts to spin. The idea of pouring concrete brought up an even worse scenario. I would have to dig to pour concrete. Digging, even in the garden, was an activity that I avoided at all costs. The diagnosis was a simple one. Writer's back. Years of sitting in a chair have given my back the flexibility of dried twigs.

"I could use cinder blocks, right?"

My lazy voice.

"You could. But the building inspector might not like that."

"Building inspector? It's small."

"Probably property and zoning as well. They inspect doghouses in your neck of the woods."

That was it. My plans were evaporating before my eyes. Not only would I have to learn how to build this shed, design a roof that wouldn't leak, and level ground on the side of a hill; I would have to do all this under the watchful eye of a professional. An inspector.

Finally I threw up my hands and voiced what had been lurking in the back of my mind. I couldn't do it. But my brother-in-law just smiled. Maybe he saw what my father had: the constant dreamer, the pie-in-the-sky element to this project; or maybe he was trying to be sympathetic. His words hit me like the concrete that I was trying to avoid.

"On the contrary," he said, "I think this project is perfect for you."

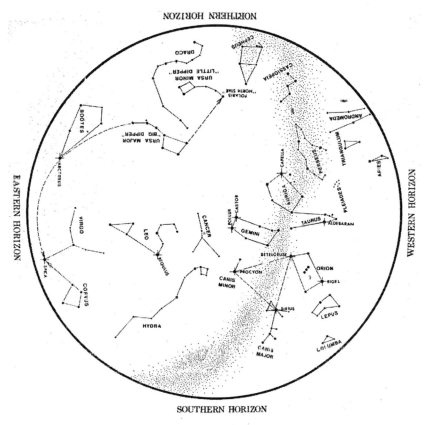

THE NIGHT SKY IN MARCH

March

Spring is in the air and will soon come. I know. At night the constellations are harbingers, reminders that the earth renews itself after a long, frozen sleep. On the eastern horizon, Leo, the great lion, yawns and stretches his lean body. It has been nearly a year since he walked out of the savannah to begin his migration westward in search of food, or perhaps to evade Orion, the hunter, who now slinks over the hill towards home, pelts in tow.

Lions are canny prey, hiding when they can, and Leo is no exception. His great, sickle-shaped head, stately and large, looms over the horizon like a backwards question mark. He looks for danger. All clear. The grasses protect his body — he blends in — and soon the full weight of the lion scampers out of the bulrushes. Leo is huge, a grown male. This constellation is easily recognizable by even the most inexperienced stargazer, with its distinctive head and body, exactly as a lion should look, and by its flagship star at the bottom of the sickle. Regulus.

Called Cor Leonis, or the Lion's Heart, Regulus is exactly that: the heart of a great beast, burning at over first magnitude. Stars were categorized by their brightness in 129 B.C., when the Greek astronomer Hipparchus produced the earliest star catalog, using a ranking system still in place today, on a descending order of scale. Sixth-magnitude stars were the faintest that Hipparchus could see with his naked eye, although an eight-inch telescope can show stars down to 14th magnitude on a dark night, and the Hubble Telescope can go deeper still, to almost 30th magnitude with long exposures. In 1856, Norman Pogson made the system more scientific by using logarithms, 2.51 per magnitude, or, in real terms, Regulus is almost two and a half times as bright as the next star in Leo, Denebola, which shines at second magnitude.

Following across the sickle to Denebola, located at the lion's hind legs, are galaxies. These faint smudges of light click off the tongue like the jerseys of a Pop Warner football team: M95, M96, M65, in the Messier catalog. But my favorite object in Leo, located near Regulus, isn't a galaxy at all but a star. R Leonis appears in binoculars or a telescope as a red ember suddenly popped from a campfire, fading or glowing with the breeze. R Leonis fades and glows as well, changing its brightness throughout the year, evaporating to a dim point, then brightening to a naked-eye object, and back again. Its cycle, like the return of life in spring, is a good reminder that everything returns to its origins, even the stars.

The constellations return too. The spring skies are woods, darker than the rest of the year, as sparse as undergrowth in a pine forest with fewer bright stars other than the eyes of night. These are animals in the sky. There is Lynx, a thin line of a constellation that dribbles, catlike, into Leo Minor, the little lion, a triangle just above Leo. Then there are the hunting dogs, Canes Venatici, running from another field to the northeast, just behind the lion, and pushing Leo into the closest stream.

Every forest has a stream, but in these woods the stream is an estuary, a saltwater marsh that hides a snake, Hydra, who slithers up from the deep south, and a crab, Cancer, next to the lion's sickle. Cancer is a loose, fan-shaped collection of stars that betrays the bony shape of fingers, not unlike a skeleton at Halloween. In the palm of the hand sits one of nature's marvels. This is the great Beehive cluster.

Visible with the naked eye, the Beehive appears as a small cloud, its stars tightly wound into a filament. A telescope or binoculars reveal more. One of the most striking clusters of the northern skies, the Beehive truly resembles bees in an angry swarm, except that these bees are stars. It is said that the ancients used the Beehive as a barometer. A keen eye is needed to see the central cluster, and this would certainly have announced a continuing high-pressure zone, bright stars, and clear skies. But I use the Beehive for something else. I use it as a marker for the ecliptic.

The ecliptic is an imaginary path, shaped like the curve of a wave, where the planets and moon roam during the course of a year. Due to the orbital path of the earth, the moon and planets remain close to this line, surfing on crest and trough over the course of time — faster for the inner planets, painfully slow for the outer ones. Once an observer establishes its invisible presence, it becomes simple to identify the planets from year to year, and to watch the moon seasonally bob up and down.

The moon is an obvious target for the beginning astronomer. New telescopes are almost always unveiled under the excitement of the moon, and the view rarely disappoints. Even through the smallest instrument, the pockmarked surface of our satellite reveals its quiet intensity; the moon is a violent place.

Throughout the eons, the moon has acted as a kind of balancing mechanism; a bar, not unlike what a tightrope walker might use, the moon rights a teetering earth. For one thing, it keeps the axis of our planet, now tilted at $23^1/_2$ degrees, fairly constant. Without the presence of our nearest neighbor, the earth itself would wobble hopelessly, like a wild drunk at Mardi Gras, lengthening our days and causing so much atmospheric upheaval that life itself would be threatened. The moon saves us. It also reminds us of the terror from space. Cataclysmic strikes from meteorites and comets have left huge craters, littering the lunar landscape with pits, fissures, rills.

On the moon, craters vie with one another as boys do at an all-girls college. They pratfall for attention, cascading into the sides of hills, pummeling the largest crater into smaller ones, clustering together with no apparent rhyme or reason. The southern hemisphere bears the worst brunt. Meteors have pulverized this part of the moon, giving our only satellite the worst beating imaginable.

Go north and the scene changes. There are more mountain ranges, hooked and jagged peaks whose shadows slink across the demarcation of light and dark, menacing the optics of every telescope with the glare of the sun. The moon is bright, stunningly so. But there are dark areas. First observed by Galileo in his little refractor and called *maria,* or seas, these dark patches were smoothed by an unknown process — whether through high-impact collisions, volcanic activity, or a combination of both, no one is quite sure. Theories about the origin of the moon are actually still sketchy. Was it captured by the earth's gravity? Or is it, as many scientists now think, a part of our own planet, blasted off in some cosmic game of eight ball?

Throughout the millennia, the earth has suffered few of the hits that the moon has taken because the earth has an atmosphere. Against a legion of angry archers, our atmosphere acts as a shield, albeit an unusual one. Most meteors burn up or are broken apart before they reach the earth, and only the very largest and most fearsome of these have made their way inside the protective cocoon of our home planet. The damage, however, has been massive.

For those of us gazing at it from the vantage point of Earth, the moon's battered face is the sculpture of an apocalyptic nightmare, a worst-case scenario of some future world gone haywire. And yet the moon is so naturally a part of our system, in symbiosis with Earth, that we easily forget its influences. As a tide maker, buffer, and romantic moll, the moon has inspired amateur astronomers for centuries, including me.

My first look through a telescope was at the moon. This was natural. It was 1966, and I was eight years old, but already I had tapped into the national zeitgeist. Space. We were going to the moon, an achievement so remarkable that it shaped an entire generation of Americans. It also shaped our technology and national pride, and it defined us, not just to ourselves as a nation, but to the very sweep of human history.

Observing the moon became my creative outlet. It was also a way of participating in the space program as a boy. On my gigantic moon map in the dining room, I followed the missions with a mad zeal, and soon the wall was dotted with pins, red, white, and blue markers where each *Apollo* had landed. It was also around this time that I decided to photograph the moon through my telescope. No easy task back then: I used my father's old Argus box camera and a steady hand.

The first few rolls were disappointing, blurred or overexposed. Finally I decided to bike to the local film developer to talk to him about my pictures. I think he was probably sympathetic to my plight, though he didn't sound like it.

"Try the library," he said. "They'll have books. Moon pictures."

He didn't understand. I wanted to do it myself.

"You'll need faster film, then."

The developer gave me two boxes of fresh film, and a month later I was back. A few of the moon pictures had turned out, and they were now hanging on my bedroom wall. But I wasn't satisfied.

"I want to do this," I said, shoving a picture in front of him.

It was a shot of the Horsehead Nebula taken by the Schmidt camera, one of the finest astronomical cameras in the world, located at Mt. Palomar Observatory. He looked at me and grinned.

"Pretty ambitious, aren't you?"

Was I still? For the past month I had experimented with taking pictures of the

Horsehead, but none of them had come out. Many had been terribly underexposed as I wrestled with exactly how long it took to take a picture. Photographs through telescopes are often long exposures, sometimes up to an hour or more, as faint light collects on film. But I wasn't shooting for that long. It was a problem that the developer had identified within me even as a kid, one that I have yet to outgrow. My lack of patience.

"To take pictures like this," he said, "you'll need to learn the craft."

He was right. It just took me thirty years to discover that.

It is significant that most observatories today are built on the summits of remote mountains. After all, these are the darkest spots for CCD cameras and photographic plates, which most modern astronomers now use, and mountains have become the best sites on Earth. But it wasn't always so. Before the invention of the telescope and even shortly thereafter, observatories were constructed in the only places possible then. In cities.

The thinking was logical and carried well into the nineteenth century. Cities were where people lived. Astronomers needed quick access to their charts and equipment, and it only made sense to build in a hustle-bustle, urban environment.

The first sites for urban observatories were cathedrals. They were tall, the modern skyscrapers of their day, towering over everything else, with sight lines that extended for miles. Many cathedrals served a dual purpose as small solar observatories useful in calculating time and dates for Easter. By the sixteenth century, cathedrals in Bologna and Rome were reliable time clocks, inspiring a new type of science: positional astronomy.

The greatest of all positional astronomers was Tycho Brahe. Born in Denmark in 1546, Brahe died just a few years before Lippershey's unveiling of the telescope. As a student, Brahe observed an unusually close conjunction of Jupiter and Saturn, noting that the astronomical tables of his day, tables calculated by none other than Copernicus himself, were off. Tycho decided to dedicate himself to astronomy, and he threw himself into his new hobby with passion.

A star would change his life. In 1572, Tycho noticed a fresh object in the sky that was so brilliant it could be seen in the daytime. Was it a new planet? Astronomers in England tried to determine whether or not it moved, but they couldn't. Tycho, who had just built a sextant not unlike the instrument that Ulūgh Beg had used a century earlier, only smaller, calculated the position of the star, and it was here that he got his idea. The future of astronomy was in careful measurements.

Tycho went on to measure the positions of more than a thousand stars, and he did this in the greatest observatory in Europe at the time, thanks to the generosity of King Frederick II of Denmark. Uraniborg, literally "castle of the sky," was a towering haven for astronomers with its sextants and transit murals. It also had the good fortune to be located in a place that is particularly modern.

On an island.

Today, one of the best observing stations in the world is in Hawaii. On top of Mauna Kea stands a bevy of telescopes from several nations, thanks to an unusual confluence of altitude, good transparency, excellent seeing, clear nights — exactly the criteria of observatory search committees everywhere.

When a site is being evaluated, astronomers rely on several factors. The first is the most obvious, culled from satellite photographs and from local meteorological and defense data: how many nights are cloudy? A good site is seldom overcast, but this is only part of a complex equation including the number of dark evenings, or what astronomers call photometric skies. Nothing kills a photographic plate like light pollution or auroras, and astronomers do their best to avoid these. Skies should be dark and transparent. They should also possess a steady atmosphere.

Called "seeing" by astronomers, the stability of the atmosphere is determined by a variety of conditions including humidity, wind velocity, and temperature variations at a site. An observatory whose thermometers move all over the place is probably one cursed with bad seeing, no matter how dark the sky is. Seeing is, in part, the result of laminar airflow over long distances, and this tends to happen most commonly around inland coastal mountains, especially on islands.

Sir Isaac Newton summed up the problem best in 1704. "The air through which we look at the stars," he wrote, "is in perpetual tremor. . . . The only remedy is a most serene and quiet air, such as may perhaps be found on the top of the highest mountain above the grosser clouds."

He was right. Valleys are avoided for observatory sites. Air gets trapped in depressions that can create the conditions for fog, dew, and haze, so the astronomer, not unlike

Icarus with his wax wings, always reaches for the sky. Mountains, being elevated, tend to win out as choices, but not every mountain is a candidate. Observatory search committees use a variety of techniques to evaluate the potency of a site, including testing with small telescopes. They count stars in the eyepiece for sky clarity, split close double stars to determine seeing, and measure atmospheric water vapor for dew and the likelihood of unexpected storms. Most sites are at least a mile above sea level, and increasing numbers of them are far from cities.

In other words, in places like Hawaii.

Even Hawaii isn't perfect. The logistical considerations of building an observatory there or anyplace remote, such as the mountains of South America, another hot spot, are legion. The availability of electric generators and fuel to power them, the cost of instrument transportation and upkeep, and local airports to fly in support personnel are all factors in determining a good site.

But for the backyard astronomer, the decisions are much easier. He has one thing going for him: where he lives. The sky there is the same sky that he sees every day, and he knows it well, clouds and all. Measuring its clarity or worthiness to astronomy isn't necessary. The amateur can't move or improve his sky; his sky just is.

I try evaluating anyway.

The ground is thick, and I leave holes in the mud and melting snow as I walk: dark pits, cavities in rotting teeth. This is my yard. The sun is shining in a cobalt-blue sky against a background of trees — there are hundreds of them, small and thick, aged and stunted. Like an astronomical Mason and Dixon, I survey, looking for the best place to build, with as much sky as possible.

The house that I live in is on the edge of a hill. We have fabulous land, provided you're a wild turkey or deer; otherwise it's mostly forest, good for animals but bad for telescopes. A fixed observatory has a specific set of rules. The quality of the site is of utmost importance. The surrounding land should be flat and unobstructed, two rules that I break on my first day of surveying. I have no flat and unobstructed yard. The best I can do is a small tract of land, scrub grass really, better suited for downhill croquet, next to the house. Hawaii or Arizona, it isn't.

"You need to build that close?"

My wife winces as I explain. It gives me the best range of sky.

"Won't our yard resemble a Monopoly board?"

All buildings, she meant. We already had a shed.

"You could use that. Cut off the roof."

"What about the mower? The kids' bikes?"

She shrugs. Most women don't consider the necessity of outdoor storage, but men do. Smitten by our garages and toolsheds, a man refuses to give away any exterior space that isn't inhabited by spiders, mice, or creepy-crawlers.

"It won't be large," I say. "Ten by ten square. Smaller, even."

"That sounds like a guard shack."

It did. I was trying to balance monstrosity and a guard shack, like most guys, I suspect, who are contemplating similar building projects. In basements and garages all across America, men are furiously building. Boats. Furniture. Our inventiveness is only constrained by our wallets and spouses, who are usually supportive.

The line that our spouses toe is a delicate one, offering tempered encouragement and love while trying to maintain their sanity. It isn't easy. Left to his own devices, a man would convert his entire home to a work-shop, with wives, kids, and dogs fending for themselves. I imagine even the architect of the Great Pyramid of Cheops, swamped at his desk with papyrus plans, listening to his wife as much as to the Pharaoh. *Just don't make it look like a Monopoly board.*

But unlike the architects of Egypt, I soon realize that I know nothing about drawing up plans.

My wife looks at my sketch of the observatory and clucks her tongue.

"I don't think they'll accept this."

The building department, she means.

"I'm just saying, won't they need to see actual plans?"

"These are actual," I retort.

One observatory, eight by eight feet square, about eight-odd feet high or so, with a roof that comes off in a fashion that I had yet to figure out. What was wrong with that?

"I think you need exact measurements."

She points to my scrawl.

"How high is this roof? In inches."

"I haven't built it yet. How would I know?"

"This isn't like writing. You can't just wing it."

They wanted real plans, she says. Numbers, specifications. Code stuff.

"What kind of code?"

She just shakes her head.

"Why don't you call them and find out for yourself?"

★ ★ ★

Building departments, I've decided, were created for people exactly like me. While many decry the regulations and the armada of rules for construction in America, I actually think they're a good idea. Safety has improved in the building industry, and so have the speed and quality of new construction, two points made to me by the clerk behind the desk.

Like many towns, the one that I live in is booming. Construction is everywhere, but not the bulldozer work of strip malls, hair salons, and fake-French bistros; ours is all homes. Many of these new homes are truly gigantic, built on incomes that I can only imagine, with five-car garages and indoor-outdoor swimming pools, guest barns, miniature soccer fields in the back. A small-town building department can easily be overwhelmed. They've seen every odd proposal to come down the pike and probably more.

"What's the strangest project to date?" I ask.

The clerk, an attractive woman who looks like she should be running a corporate board meeting instead of talking to me, tugs on her tortoiseshell glasses.

"Well, yours is pretty unusual. An observatory?"

110

Lest she think I've just won the lottery or am a famous actor, I explain the project to her. More like a cheap garden shed, really, than what was probably in her imagination, an expensive and mammoth gold dome built atop my sprawling country estate. Sheds, I discover, are also under the auspices of the building department.

"You're doing this yourself, then?"

I am and I produce a drawing to prove it. She smiles.

"We'll want a little more detail. See here."

She produces a sudden ream of paper containing all the specifications required by law. Types of lumber, beam centers, grades and angles. She tells me that I don't need an architect but to merely do the best job I can by sketching it out. Then she hands me a pile of papers to be signed, notarized, paid for.

"Signatures at the proper department and return. Good luck."

Driving home, I realize that I have drawn only one architectural plan in my life. I was in the eighth grade, and it was a class in drafting. I hated it, and all I remember is sitting at one of those angled desktops, scribbling away. But mostly I recall how difficult it was to render a three-dimensional object on paper. Impossible. I'm no artist. But I'm not a builder, either.

After drafting, we went straight to Mr. Geiger's woodworking class. Geiger was a perfectionist, a man who took pride in his craft and expected his class to do the same.

"Gentlemen, you've dreamed. Now it's time to build."

My project was an electric lamp. It worked, barely, though I never could get the balance right and it leaned perilously to one side, threatening another Chicago fire. Geiger gave me a D. The only D given in wood shop class that year.

Maybe it's thinking about my D in wood shop or the idea of a spectacular disaster, but a thought strikes me. The roof. I won't pull it off in one section, I decide, counter to the ease that Mr. Geiger would no doubt suggest; instead I'll split it in two. The motion would be elegant, lightweight.

It even wins approval from my wife.

"It's weird, but I like it."

"Then I can do it?"

She smiles. This is the secret of marriage. Dreams are balloons, and our partners provide the wind. Of course I can build it.

"One question. What are you going to do out there all night?"

★ ★ ★

Most people think that astronomers flirt from star to star, like Casanova, or worse, that they just flog the same few objects in their telescopes, hoping for something to happen. Not quite. Many amateurs have carefully structured observing programs, backyard science that they hope to accomplish under the stars. My interest is Jupiter. While some stargazers lust after comets and deep-sky objects, I find myself drawn to the largest planet in our solar system, and for reasons opposite to those of most observers: because it changes.

As astronomers, we never see the surface of Jupiter. Instead, we observe only the tops of its clouds, a swirling mixture of hydrogen, ammonia, and other gaseous elements in movements choreographed by a rapidly moving planet that displays spots, festoons, and dark bridges across our telescopic view. Some materials move faster than others, and a careful amateur can time this, helping science by establishing the various rotational drifts of the many currents.

Jupiter is a planet of mysteries, both on its surface and in orbit, with its plethora of moons, though only the four brightest are visible through small telescopes. Discovered in 1610, these moons, aptly named the Galilean satellites, perform a nightly dance. Like runners at the Olympics, the moons scramble for the inside track by weaving in and out of position — sometimes close to the planet, sometimes far — revealing the hidden secret of the cosmos.

Nature is very busy.

Io, the brightest innermost satellite, is perhaps the busiest of all the moons. One of the most volcanically active objects in the solar system, Io is a bouncing infant after dinner, spewing its volcanic meal into space whenever it turns. None of this can be seen with a small backyard telescope, of course, though an observer can note albedo, or changes in brightness and contrast, as Io makes its way, every 1.76 days, around the gas giant.

Europa, the sixth-closest moon to Jupiter, is a different story altogether. The surface of Europa appears frozen, and scientists have been debating about whether there is an ocean beneath that ice, one perhaps capable of sustaining life.

The question of life besides that on Earth existing in our solar system has always been a fascinating one, and it is probably the first question posed to stargazers by the public. "Have you seen a UFO?" Interestingly enough, few UFO surveys focus on backyard astronomers — an irony, since we spend the most time watching the skies. But just as well. Astronomers might offer their actual opinion: that life very well might teem throughout the universe and we just haven't seen it yet.

Recently, newspapers have claimed that life has been found in Martian meteorites that struck Earth, tiny fossilized microbes of a remnant past. The jury is still out. Arguments continue, as they always do, but scientific arguments can easily turn on themselves. We know that tiny meteors hit this planet all the time. Is it possible, then, that pieces of Earth, with our own native bacteria aboard, have been jettisoned into space and have reached other planets or their satellites? The supposition is tantalizing: perhaps right at this minute, earthly microbes are stirring under the ice of Europa.

Mysteries abound on Jupiter itself, but none is more mysterious than that surrounding the Great Red Spot. Like the blinking eye of a giant Cyclops, the Great Red Spot hovers and weaves it way around the planet, wandering in longitude. Otherwise it is rather fixed. Several times wider than the earth, this great oval appears to be a whirlpool or vortex, not much different from our terrestrial hurricanes, except for this. It has been around, month after month, rotation after rotation, for almost three hundred years.

My last look at the Great Red Spot was long ago, right after high school, when the spot was dark, obvious even in a small telescope, like a black eye in need of a cold compress. Now I wasn't so sure. According to calculations in *Sky & Telescope*, the monthly bible of amateur astronomy, the Great Red Spot was supposed to cross the central meridian, or the imaginary line dividing the north/south center of the planet, at any moment. But all I was seeing was an empty shell, a colorless impostor.

Was it me? I rubbed my eyes. The floaters and retinal gunk had increased over the years, canoeing over bright surfaces and making them gray with spiderwebs. Observing the moon has become an aerobic exercise, flipping and rolling my eyeball to clear out the floaters. It doesn't work. My eyes are shot. One, a victim to corneal keratoconus, refuses to focus; the other, my trained astronomical eye, prey to floating crud.

A beginner will often peer into a telescope for a few seconds and proclaim confidently that he can't see anything. But a seasoned observer knows better. The atmosphere is constantly moving. It can vary from an image that simply hangs stationary in the eyepiece, as a painting does on an artist's wall, or it may ripple violently like the writing on a newspaper held behind a running faucet. An astronomer has to be patient then, like a good hunter. He has to wait at the eyepiece and pick his moments, training his eye to locate the faintest hint of detail, a process that can take time to develop.

Time doesn't help me. Every glance at Jupiter reveals the same thing: a border marking the thin remains of where the Great Red Spot should be. Near the edge of a darkened South Equatorial belt, I can see a faint hollow or depression, like a dimple of a golf ball, but nowhere do I see the dramatic color that I remember from my youth. The Great Red Spot appears to have become the Little Pale Blemish.

The universe changes, but what rarely changes is our reaction to it. Creatures of habit, we are shocked. The world is moving on without us, we cry, and history stumbles along with it. Or maybe it's simply nostalgia. Nostalgia is part and parcel of the human condition, a natural gift of aging, and it protects us from the message that the universe telegraphs to us with frightening accuracy.

You are small and will soon be forgotten.

Observing the Great Red Spot after such a long layoff has sent another message to me: I am behind the curve. Modern astronomy has passed me by, and I am merely recycling my old knowledge, information, and telescope tricks, permanently frozen in 1976. Nothing in my head has expanded since then. In fact, my astronomical knowledge is contracting, and the plummeting pull that I feel isn't the collapse of my own gravitational star, it's the realization of how little I know about a hobby that I love.

Spring

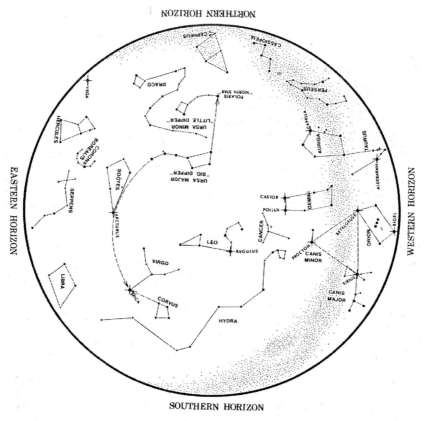

SOUTHERN HORIZON

THE NIGHT SKY IN APRIL

120

April

Twilight drops like a slow curtain. On the sunlit edge, a band of fruit colors — citrus orange, strawberry, melon, and peach — gather at the horizon, brushstrokes that grow darker toward the night, which is already here, claiming its place. The stars above me are beginning to show their faces as well, with Betelgeuse in the west and Arcturus in the east, a bride and bridegroom exchanging quick vows before one of them leaves for a year.

Betelgeuse will exit first. He dips under a peach band, and in a few weeks the motion of the earth will claim him until another season, but tonight I witness his silent retreat. The field that I am standing in, less than a mile from my house, opens up to sky on all sides of me and I can see forever. Why I stand here is obvious. There's a comet in the April skies.

Comets have, since the beginning of human time, inspired and frightened. They were viewed as ominous portents, bad karma, purveyors of rotten luck. Kings died under the appearance of hairy stars, or lances in the sky, as comets were called. Crops failed when they showed up. Cities were sacked. Pliny, the Roman author, summed up best the events of 48 B.C. when he wrote:

> We have in the war between Caesar and Pompey an example of the terrible effects which follow the apparition of a comet. . . . That fearful star, which overthrows the powers of earth, showed its terrible locks.

In 79 A.D., a comet was said to foretell the death of the Roman emperor Vespasian; in 455 A.D., the actual sighting of one was enough for Valentinian. It was a sign that he knew too well: a leader must go. He obliged, dying soon afterward. Comets have been blamed for the defeat of Attila the Hun at Chalons, and for the battle of Hastings, but seldom are they mentioned as the awe-inspiring events that they really are.

Once the stuff of legend, comets are nothing more than dirty snowballs of ice, rock, and ionized gas that ignite as they approach the sun. Every comet has its own path, not unlike people. Some move in great elliptical sweeps, leaving their homes, like phosphorescent salmon, only to return years or eons later, in glory.

Some glory, however, is outlived. Comets can become duds. Lubos Kohoutek discovered this when one of the brightest objects ever seen from Earth fizzled. Comet Kohoutek became a metaphor for failed promise, wasted youth. But not all comets are so. Others, like Hale-Bopp, which I saw, and Hyakutake, which I did not, were great public events, naked-eye comets with tails that stood on end as short hair does after a shower.

Comets also offer a glimpse of immortality. Beyond their primordial gases and early solar-system materials, the discovery of one of these objects gives the hopeful backyard astronomer a rendezvous with history. He might someday give his name to a comet. Who hasn't heard of Halley?

All over the world, on any given night, comet hunters search methodically along the morning and twilight horizons just after the sun sets or before it rises. The hunters move their telescopes and large binoculars in straight, linear searches. Comets are often found near the sun, and they can escape the eyes of many larger, earthbound instruments, though deep-space probes may soon horn in on the action. The odds are stacked against the comet hunter, like the odds of finding gold in a Connecticut stream. Thousands of hours, even years, are spent with no guarantee of success.

Why do it?

Most enjoy the hunt. Others use comet searching as a way to increase their knowledge of the night sky. If you want to find a source familiar with the stars, don't consult a computer or a chart. Talk to a comet hunter. They know the sky well. Most do it out of love, I suspect, and not for the glory of the find, even though the find can be glorious.

Just ask David Levy, the codiscoverer of Shoemaker-Levy 9, a comet that nobody ever saw visually. It was too faint. But most folks heard about it. This was the object that slammed into Jupiter in 1994, the pictures of which were beamed around the world. Jupiter took these hits, a champion prizefighter on the ropes, night after night without falling. The scars remained for months after the assault, black spots floating in the high Jovian atmosphere, but the memory of the beating lives on, inspiring a new scientific pursuit that involves the precise plotting of near-Earth objects.

Suddenly astronomers saw with their own

eyes the destruction that could be wreaked by objects from space, and all over the world, telescopes both amateur and professional were mobilized. Many observers looked for rogue comets, but most instruments were pointed to our nearest threats as well. Asteroids. These chunks of iron and rock, some tiny, others larger than football fields, hurtle through reckless swarms of debris by the gross. There are countless thousands of them, darting around the sun in all directions imaginable. Their orbits used to be like the talk of braggart gunslingers in the Old West: many of them unknown and untested. No longer. Thanks to Shoemaker and Levy, we are all watching now.

But not all comets are so dangerous. The beauty of a newly discovered comet, particularly one visible to the naked eye, is rarely forgotten, not unlike an old girlfriend whose face remains burned in the memory long after she has gone. My first great comet was West, in 1976, and I remember dragging my mother out of bed at 4 a.m. for a look. She was suspicious of all comets, of course, but bright ones made her skin crawl as she recounted their tainted history. Comets were the jokers in her astrological hand. They announced that something unexpected was going to happen, good or bad; harbingers of change that come out of nowhere.

"Never, fear change," she said. "It's natural."

But I wondered if she believed that. She walked out of the house clutching her robe as though it were a talisman sewn to protect her.

Her first look at West rattled her. The comet was bright, a blue welder's arc that revealed the fan of a tail, or several tails, spraying away from the horizon. She crept up to the telescope with a mixture of anticipation and doubt, pausing for a moment before looking.

"Stunning," she said, watching it with her naked eye. "No wonder the ancients were afraid. Many felt you couldn't toss a stone in the air without inspiring a hurricane, and this is some hurricane."

There was a kind of Zen quantumness to her thinking: the belief that everything in the universe is linked. Science might not argue with her today as passionately as I did back then.

"That's foolish, Mom. Small events can't shape larger ones."

"Why not?"

"A connection suggests meaning, even consciousness."

"And the universe isn't —"

"Sentient? No."

She smiled and leaned in to the eyepiece. The next sound that I heard was a gasp, as though she had just seen the most beautiful painting in the world, nature's painting, on the canvas of night.

"Tell me that doesn't look alive," she said.

It did. There were glowing filaments to West, streamers that ran from the nucleus, at the head of the comet, now split and shattered into fragments, all the way to the tail. My mother must have stared into that eyepiece for twenty minutes, longer than ever before, and she kept watching even as dawn began to approach. Then she turned to me.

"I think just seeing this comet is enough to change lives, don't you?"

It was enough to change mine.

Now there was a new comet in the April skies. It was also bright and easily seen. Though I hadn't observed a comet in years, I felt my spine tingle with excitement. Every comet, like every child born on Earth, is distinct. There are no identical copies, and no two act the same way. Comets express their free will by brightening or dimming, often unexpectedly, according to some unknown whim.

This one was no different. Heralded as the best naked-eye object in the past five years, Ikeya-Zhang, as it was called, was brightening, and fast. Discovered a month earlier by two amateur astronomers, Daquing Zhang from China and Kaoru Ikeya, the internationally known comet hunter from Japan, Ikeya-Zhang was offering a perfect opportunity for photography.

For the past three months, I had struggled with taking pictures the way most beginning astrophotographers do, learning on the fly. Now it was time to deliver. At Ikeya-Zhang's brightest approach, just days before its picturesque encounter with the Andromeda galaxy, I set up my Canon Elan in the field close to my house. Several rolls were shot from a tripod first — simple enough, for the rotation of the earth won't get in the way with such short exposures — and they came out well.

Success can embolden a person, and it certainly did me. Unsatisfied with my wide-field shots, I decided that I wanted the classic telescope picture, gossamer and romantic, straight from the colorful pages of *Astronomy* magazine, even though I didn't know what I was doing. The pictures proved it. All I got back was fuzz.

Had my comet moved?

Not exactly. In my haste, I'd made the mistake of many beginning astrophotographers, particularly those standing in open fields. I didn't spend enough time aligning my telescope. An unaligned telescope is like a trailer hitched with loose chains to a car; its wiggles and gyrations invite disaster on film. The remedy is a simple one. Aim at Polaris.

The North Star, Polaris, can be found by every woodsman, Boy Scout, and camper across America, but the constellation it resides in, Ursa Minor, often cannot. There is good reason for this. The stars around Ursa Minor often appear muted from the suburbs, and if not for the open bowl of Ursa Major herself, which points to Polaris, most folks would be lost. Where is north?

In the field, north is obvious. High up, unobstructed by trees, sits Ursa Major, the Great Bear. Known to everyone, backyard astronomer or not, as the Big Dipper, Ursa Major is probably the most famous marker in the sky. Its seven bright stars are burned into our minds as children, usually as the first constellation that we can identify, and one that carries us throughout our lives. No matter how far we tread in this hemisphere,

Ursa Major is there, pointing us home. If south is Oz, uncertain and fanciful, then north is our Kansas.

But unlike Kansas, many of the sights in Ursa Major are otherworldly. There are faint planetary nebulae and galaxies by the fistful, some of them bright and well developed. The most famous of these, M81 and M82, arrive as a pair, strange as the theater's Odd Couple; these galaxies are easily located to the north of Dubhe, the brightest star in the bowl of the Dipper.

With a small telescope we can see their quirks. M82, the cigar-chomping sportswriter Oscar Madison, really looks like a badly rolled cigar, thin and washed-out; and M81, the elegant, tightly wound Felix Unger, is a spiral galaxy of such neatness that it demands a second glance. They are inextricably linked in the eyepiece, occupying the same low-power field, a field that displays the humor of the universe.

Humor abounds in Ursa Major, though. At the handle of the Dipper is the seventh-brightest star in the constellation, Eta Ursa Majoris, or Alkaid. Follow Alkaid east and the next stop is Arcturus, the fourth-brightest star in the heavens. Go west, up the handle, and the comedy routine continues. The star is Zeta Ursa Majoris, an object that teases the imagination: is that one star I'm seeing, or two?

The Arabs have other names for Zeta Ursa Majoris. Alcor and Mizar, the horse and rider; or Al Sadak, the test. For countless generations, eyesight has been measured by Mizar and Alcor, two stars, close in brightness, that are separated by only a thread. Through a telescope the separation is obvious, with a closer second pair, though few can see the first separation without optics. There are reports from fourteenth-century Arabia that people lined up to test their visual acuity, like reading an eye chart before any existed. It is a test that I fail miserably. I've never seen the two split, but then, I sometimes have a difficult time just making sense of the constellations.

Looking up at Ursa Major makes me shake my head. What were the ancients thinking? Nearly every culture from the Greeks and Romans all the way to the Iroquois and Algonquin has identified our Big Dipper as a bear. How did they eat? Wasn't it obvious to them that this was a ladle, a scooper, a spoon?

Modern Western conveniences aside, I'm sure the real reason that I'm unable to locate a bear within the confines of the Big Dipper is simply because of light pollution. The creeping glow of cities and backyard porch lights has overwhelmed many of the surrounding stars around Ursa Major, leaving little more than the famous seven. But on a star chart or through binoculars I must concede my ignorance. Ursa Major really does look like a stalking animal.

From the bottom and west of the bowl are two hind legs, arched and thundering through the woods. The tail is up, balanced by the stars in the dipper handle. In the opposite direction, the bowl now appears as the torso of the bear, with a triangle-shaped collection of stars for the telltale pointy skull. This bear is large, angry.

This bear is also a mother.

Go north from the brightest star in the bowl. This is Polaris, the North Star. But Polaris betrays its own handle, a replica of Ursa Major in miniature. The stars of the two dippers could easily be stacked inside each other as pots in a kitchen are. The ancients saw these two in tandem, as a mother and her child. Ursa Minor, then, is the little bear.

Cubs find trouble and this one is no different. To the north of Ursa Minor, slithering its way between cub and mother, is Draco, the dragon, a line of stars that lean backward, like a snapping reptile. The

Egyptians saw Draco as a Nile crocodile, and indeed, it does resemble something primeval with a pointed snout, its jaws pulled back and ready to strike.

Danger abounds in the spring forest. In no doubt his first time away from his mother, the cub encounters a horrible dragon. It's the fear of all parents when they send their children out into the world, though few voice it. Is the child prepared? The little cub is. He's instinctively holed up somewhere, a cave perhaps, but Draco is clever. He waits. He has that luxury, waiting.

Looking at the bear cub that night with the comet sinking in the west, I realized my problem. If I am to align my telescope so that its electrical drive matches the rotation of the earth, the axis of the telescope must be pointed not just to Polaris but to a position less than one degree northeast of there to the North Celestial Pole. Anything less accurate and the field moves.

Correcting the issue could have helped, but it didn't. By the time I had the telescope properly aligned, the comet had veered behind a low stand of maples. It was too late. And when it rained the next week, obliterating my chances for a great picture, the comet was gone. Behind the sun. Ikeya-Zhang will return again, I know, not in my own lifetime but in the lifetime of my distant progeny — hundreds of years too late for me.

Galileo, the first man to use a telescope to scan the heavens, never built an observatory. He didn't have to. Peering out the window of his house in Padua provided him with all the stars that he needed. His telescope was also small. But as instrumentation grew larger, and longer, a need arose for separate buildings, and they were often connected to local universities. The first of these was in Holland, at the University of Leiden, where a dedicated astronomical facility was built in 1632.

By 1673, a famous Dutch telescope maker by the name of Johannes Hevelius began to construct aerial telescopes that ranged up to half a football field long and were manipulated by a series of masts, pulleys, and shipping ropes. These needed buildings to house them, permanent stations or observatories, and a small army to operate them properly. Hevelius knew the problem right away. "The enterprise," he wrote, "is not one for a private person to undertake, but for some gentle nobleman possessing ample room, money, and above all, enthusiasm."

In other words, only a wealthy amateur could afford an observatory. Fortunately, there were kings and governments. Thanks to Louis XIV, the Paris observatory was constructed in 1667 as a kind of clearing-house for matters scientific. Chemical experiments, mechanical models, and natural-history collections all shared a roof with the astronomy department, not unlike a modern university. A rectangular structure with two octagonal towers flanking its sides, the Paris observatory employed a few of the architectural tricks gleaned from the old Islamic observatories, notably the addition of offices, a lecture hall, and permanent living quarters for the astronomers. The observatory also housed several aerial telescopes, a tubeless instrument on the roof, and a 120-foot behemoth located on the east tower.

Not to be outdone by the French, the English King Charles II issued a decree: "In order to the finding out of the longitude of places for perfecting navigation and astronomy, we have resolved to build a small observatory within our park at Greenwich." This became the Royal Greenwich Observatory.

The first royal "astronomer-observer" was the Reverend John Flamsteed. In 1675, Flamsteed, who possessed no formal astronomical training, created a mathematical model for placing celestial objects by exact coordinates. This method is called right ascension and declination, and is still used today. He also cataloged stars, making the facility in Greenwich one of the most famous observatories in the world.

By the early eighteenth century, astronomical observatories were popping up all over Europe. Many were rooftop affairs hitched to great towers, some rising above the buildings that served as their foundations, others just rising straight up from the ground. These tower observatories spawned imitations and constant, if sometimes eccentric, refinements. There were large, open windows for airflow, and a few of the towers used plants and trees, serving as mininurseries that the astronomers hoped would stabilize the image in the telescope by sucking out all the bad and tremulous air.

Early builders also began to employ the amenities that contemporary observatories all over the world now use. This began with the Swedish astronomer Anders Celsius. In 1740, while remodeling his house for use as a home observatory, Celsius came up with an innovative idea. On the roof he designed a rotunda or dome to shelter his instruments, bringing the sky closer to earth and no doubt saving him on runaway construction costs.

Soon domes were part of the observatory planning process, and by 1785, with the building of the Dunsink Observatory in Ireland, the visual vocabulary was instantly recognizable. Rotating domes would now grace observatory roofs everywhere, and along with the innovations of libraries, offices, and residences on astronomical campuses, the professional design of choice had been set.

Not for my observatory.

In home building stores, there are plans for constructing outdoor gazebos, birdhouses, and log swing sets for children, but no plans for home observatories. There is good reason for this. A telescope shelter is highly personal, subject to the eccentricities of the owner and often difficult to copy, though I tried my best. Like many of my better ideas, the observatory design that I submitted to the building department in town wasn't original. It came from a 1951 photograph by British astronomer J. Hedley Robinson, who built a split-roof observatory for his five-inch refractor in his own backyard. The walls and roof (consisting of a flat, sliding sheet of transparent plastic) were constructed cheaply, at least from the looks of the picture, but there was a stoic simplicity to it.

Robinson's observatory was light and easy to build, prerequisites of mine. I had a single set of hands, and two more pairs if I counted my elementary-school daughters. Weight was important. I wanted the easiest roof I could make, and I cribbed Robinson's idea with all the verve of an escaping bank robber.

Most backyard astronomers employ a full roof, not two separate halves, for their observatories. They use wheels to push the top off its moorings and to keep it there, resting the roof on large beams to prevent disaster. This accomplishes several things. It is simple to build, and it guarantees that rain and snow will stay out, two points that seemed to elude me. When a roof splits in half, it doubles the amount of area where water can escape and triples potential problems. Rainwater and melting snow will find a way into any crack, and often do. A split roof only complicates matters. It has to stay closed in inclement weather, surviving thunderstorms, potential downdrafts, and blizzards. It also had to stay attached to the shed, a trick that I couldn't figure out.

"Your roof has to be captive," Jim said.

Exactly the way that I was feeling about this project.

"Secured, I mean. Think hurricanes, nor'easters."

"What will do that?"

"Screws."

He grinned. The irony was clear. If I screwed the roof down, then it wouldn't open. And an observatory without sky was just a garden shed.

Inspiration, I've found, arrives at odd moments. I got mine while pulling into the garage a second time and watching the door lower. There were wheels that opened the garage door, moving on tracks, and this combination, I figured, could easily suit my needs. The paradox of garage door design and observatory roofs is similar. Both must move with a canny flexibility, yet be secure enough to survive fierce weather.

My brother-in-law nodded.

"Good idea. Now what are you going to do about your hill?"

My hill. For the past two weeks, in weather more like that on the Irish coast, wet and bone-cold, I have been outside, moving a great pile of mud from my neighbor's forest into my own. The site that I had chosen, next to the house, had to be cleared and leveled before any building could start. I gave this project a Saturday; it took a month.

Since I promised my wife that I would build the observatory as cheaply as possible, my vow of poverty precluded several creature comforts. For one, I would clear my own land. This meant the hoisting of rocks, miniature boulders just under the topsoil where I needed to dig my pier. I would also haul my own soil; or rather, I would borrow my neighbor's dirt left over from an underground electrical project and engineer it over to a second, growing heap that I hoped to level. A bulldozer could have done this in a few hours, but bulldozers cost. I, on the other hand, worked for free.

None of it seemed very thought-provoking. I had a two-foot dip over ten feet before a steep, perilous descent down a ravine. What was two feet?

Rock, I discovered, was everywhere. In Connecticut, we are all living on a gigantic slab of bedrock with only a dash of topsoil. Farmers, during the founding of America, used this land to grow food, an idea made only more amazing when you actually begin to dig. Two inches and the blade stops. Moving doesn't help any.

Stone occupies a large portion of my yard. There are boulders and antique walls, and rocks pop up in the unlikeliest of places as do weeds in a garden. They are also heavy. More than two hundred years ago, farmers in Connecticut concocted an idea bordering on genius. They would build stone walls. These walls served as property demarcations, fences for livestock, and erosion breaks. But just as important, the walls got rid of stones from the fields.

Someone, a Yankee Albert Einstein lost to history, came up with the concept. A simple stone wall, he reasoned, measuring three by three feet would displace more stone than simply chucking it on a pile. And so the walls were built, testaments to time and hard work before any of us were born. Fields were also cleared, and except in the forest, where I live, the digging became easier.

But not in my yard. I begin with medieval digging tools: a hammer, a straight iron, and a spade. I also use the geometry of the Egyptians, the mathematics of Stonehenge — twine, sun, and sticks — to lay out the observatory.

My wife watches the progress. A businesswoman, she's used to efficiency and I wasn't even close. My work is Stone Age; slow and laborious, it's the labor of a man who has nothing better to do with his time.

"Call in a truck," she says. "Get some more dirt."

She looks at me sadly. My face and clothes are filthy. I'm walking like a penguin with scoliosis, a crimped waddle because of my aching back. Already I had dumped, by hand, fifty loads from the top and rear of the yard, with no end in sight.

"Don't chintz. A foundation's important, you know."

Building, like writing and cooking, seems to invite criticism. People who have never picked up a hammer before, and even some who have, suddenly begin offering me their advice. I shouldn't be surprised. We all build. Like a succession of Legos, each one clipped to the next, our lives are a construction of who we are. Some lives are well conceived, others less so.

Construction happens this way too. Blueprints are drawn, but an inevitability always pops up. The unplanned. The ghost in the machine. A good builder is flexible. Looking at my small pile of mud that evening, I knew one thing: the observatory would be built entirely by me. No bulldozers, no trucks of soil, no last-minute pleas to the carpenter up the road for help. This was my project, and if it took a miracle for me to finish it, then that's exactly what I would expect. A miracle.

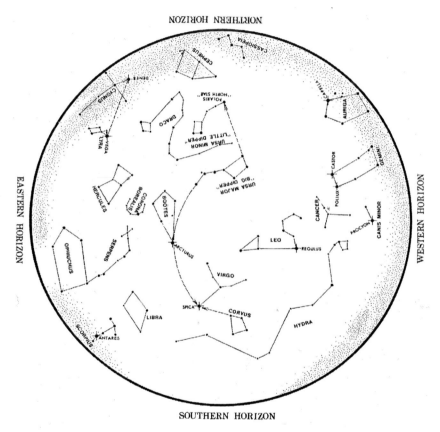

SOUTHERN HORIZON

THE NIGHT SKY IN MAY

May

Spring also carries miracles, and nobody knows this better than the farmer. After a long and dormant winter, the ground is renewing itself, greening up, and the stars above reflect this. Overhead there is the lion, but Leo isn't the symbol of spring, the spring of farmers and planting crops; that duty belongs to Virgo.

Virgo, the maid of the wheat field, is perhaps the most ancient of known constellations and yet the least disputed. Nearly every culture on earth sees Virgo in a similar way. In India she is Kanya, the maiden, mother of Krishna, who carries in her bosom the fruits of the earth, and in her hand a sprig of wheat. Wheat is everywhere in the Virgo story. Isis. Bethulah. Daughter of the fields.

In the sky, Virgo looks like a stick drawing made by young children. Located to the southeast of Leo are the appendages of Virgo, splayed and dangling in postures seldom seen in real life. Her legs are bowed, as though playing tunnel with toddlers. She uses her arms for balance, or maybe it's a welcoming hug. Virgo welcomes spring with her presence, a sign first found in the rising of Spica.

The sixteenth-brightest star in the heavens, Spica partners with two other stars of the spring triangle: Regulus, at the leg of Leo, and Arcturus, the coal-red ember burning in Boötes. Spica is said to represent the ear of wheat extending from the left hand of the maiden, down from her kite-shaped head. The constellation, in my eyes, is a bit of a stretch for such a beautiful woman. The stars of the head, which is too large for a real human being, are faint, and though well established, they seem to linger longer than necessary, like the manes of rock guitarists, which is perhaps the point. The maiden is all hair and legs.

But in a telescope, Virgo is more than just a pretty face. She has the gaze of a mystic, a seer. Her long body, thinned by the planting, allows a medium-sized tele-scope into the deepest parts of her soul — the entire universe. Virgo is pointed not to-wards our galactic center but away from it. Therefore, many of her secrets are of places outside our own galaxy. Deep sky rules in Virgo. It comes in the form of a swarming nest of objects, the famous Virgo cluster of galaxies.

If studying the heavens heightens our sense of insignificance, then the Virgo cluster confirms our worst fears. The earth is a grain among a myriad of grains. Perhaps it's worse than that. The Milky Way, our spiral galaxy that is home to all the stars that we see and many of the deep-sky objects as well, is just a speck among countless specks, and in Virgo the specks turn into a small Jackson Pollock painting — more splatter than the eye can take in.

These tiny specks are blurs in a telescope — island universes, a term coined by the German philosopher Immanuel Kant in 1755 and still apropos today. Man may never leave our own Milky Way galaxy. The distances are vast, and we seem marooned in an ocean so large that its size is almost incomprehensible. The Andromeda galaxy, our closest major galactic neighbor, is two million light-years away, or, to put it in the vernacular, it took light two million years at the breakneck speed of 186,282 miles per second to reach us. And Andromeda is just another speck.

But a backyard stargazer, looking up, doesn't feel this desolation. He sees the wonder of creation, like a spiritualist, and also the hunt, like a pirate. The Virgo cluster is priceless quarry in the telescope. Many of the galaxies are dim, especially when seen through modern, light-polluted skies, and they are notoriously difficult with small-aperture telescopes, a fact that I quickly learn.

The Virgos stretch out, in a circular patch, just southeast of Denebola, the tail of Leo. For a week, the Virgos have tormented me, not because I didn't know where they were, but because I couldn't see them. Darkness is important when trying to find faint fuzzies, and properly adapted eyes are a must. Eyes spent in darkness for an hour gain a sensitivity that television eyes do not, and an observer needs every edge. But my house isn't dark enough. There is light leaking from every window, it seems, casting soft shadows, even though half the family is supposed to be sleeping.

"Off with the reading lamp," I yell up to my daughter's room.

"It *is* off," she says out the window.

Like those of many young children, my oldest daughter's room glows at night like a carnival midway. There is a bevy of buzzing battery-powered and electric devices, and diodes from games flickering against the wall, blue star lights, clocks and dolls whose eyes light up when they talk. It isn't enough just to pull the plug on her room; I have to pull the plug on American pop culture, or at least throw a blanket over it.

The window slides back farther. "Everything's off now."

I look up. Her room is actually black.

"It's dark. Can you see out there, Daddy?"

I could. In the eyepiece, the first few wisps come into view, pencils of light, and then the elliptical smear of M87, an oddly shaped galaxy in Virgo. M87 is a strange duck, emitting a jet plume of light from its center, like a branch, straight out. X-rays and strong radio emissions have been detected emanating from this plume, suggesting that it may be material vented off by something more mysterious: a black hole.

Black holes are said to be the result of massive collapsing stars. Gravity keeps on going, pulling everything in its wake, including light. Nothing escapes the clutches of a black hole, except perhaps a child's imagination.

My daughter is still at the window.

"Are those stars or planets?"

She points through the trees, counting four.

"Planets."

One's missing though: Mercury, which had set an hour earlier.

"That's half the solar system," she says. "Isn't that weird?"

It was.

The planets were clumped together that night, a bright row of headlamps on nature's highway. Not very unusual, but it was rare enough to warrant press coverage. An historic gathering of the five major planets, cried the newspapers, and happening now.

Since the dawn of human history, people have looked at the night sky with a mixture of awe and foreboding — awe at the enormity of the universe and our smallness in it, and foreboding for the exact same reasons.

The equation used to be a balanced one, equal parts respect and fear, but in my seven months since getting back into astronomy I had noticed that fear had seized the higher ground. People were nervous, uncertain. And the alignment of the five brightest planets in the sky, all grouped together within a tight pack just after dark, seemed to heighten this mood.

The night sky has long been a magnet for the beliefs and superstitions that humans concoct, covering everything from alien influences to faces floating in the clouds. The first cave dwellers, ignorant and fearful, drew pictures of meteor showers on cave walls, only they didn't know that these were rocks from space rather than pieces of the sky falling. Science had to take a backseat to human interpretation — which held that the cosmos dictated our lives — and it has been fighting back ever since.

To the animist, tree fairies, gnomes, and unseen forces are as responsible for natural phenomena as any scientific explanation, and their explanation can seem just as irrefutable. When a star falls from the sky, the animist might believe that it was thrown by the keeper of the star, whereas the scientist knows that it is simply a rock hurtling to Earth at random.

As I looked at the planetary gathering that May evening, similar thoughts came to my mind. What were the planets telling us about the near future? And why should we listen?

Growing up in the household of an astrologer had made these questions all the more pertinent. Earth and sky were simply mirrors, my mother said, of the inner or political conflicts in the world.

"Are we that important?" I asked. "People?"

"Of course."

"But you don't feel that we were just plunked here?"

I evoked the angry mathematics of a teenager. X number of sustainable planets multiplied by near infinity came up with a number too large to wrestle with, clearly a dice throw. Why should the universe possibly care about us?

"Because we're alive," she said.

"So too are protozoa and mice. Do they have astrological charts?"

"A chart that would only matter to another mouse," my mother joked.

Her voice was canny. The perspective of the observer, she said, had everything to do with how we interpret the world.

"Meaning?"

"If you were a mouse, there would be mouse signs."

Thirty years later, the signs were there, for mice and men. The gentle foreboding that I'm surrounded by, first from the comet, and now from the visual gathering of planets, seems to fuel overactive imaginations. A quick glance at the newspaper says it all: bad news. Wasn't a cosmic commentary almost expected, if not demanded?

But the universe of the backyard astronomer is, at best, a recycled commentary. The planets have aligned themselves countless times before, and they will again, with or without us. Even comets come and go with clockwork regularity, but what remains fixed is something ancient and terribly human: the need for people to connect with something outside themselves. Cosmic signs, then, represent our desire that the universe, the world that swirls around us, actually cares. They are an investment in our failing hopes and rising fears as a society and as individuals: a hope that when we look up at the heavens, the heavens are busy looking back. And maybe they are.

The stars at night comfort us with reliability. For generations, people have looked up to wonder, argue, and plead. They have done this from caves, in the backs of pickups and on the backs of horses, in fields and on patios, in mansions and in huts. The stars demand our awe and respect, but they also require something else: that we look up. "Some things are never clear," writes Robert Frost, "but the weather is clear tonight." And there is no better call to arms for the backyard stargazer. A clearing sky, stars on the rise. The signs can take care of themselves.

They always have.

The first American observatory, in a sense, wasn't one at all. Surveying stations came of age in 1763 with the arrival of two men from England, Jeremiah Dixon and Charles Mason. In the early days of the colonies, astronomical instruments were rare, imported from abroad and used as common tools. Many of these were not telescopes at all but transit instruments, useful for determining geographical coordinates and helping to resolve land disputes. One of these disputes, a nasty border spate between Pennsylvania and Maryland, drew in Jeremiah Dixon, a master surveyor, and his astronomer accomplice, Charles Mason.

Mason and Dixon created three observing stations in Philadelphia and its environs, and then used a telescope to measure the eclipsing moons of Jupiter to determine terrestrial longitudes. But the stations weren't meant to be permanent, and they were quickly broken down.

It took a master horologist to take this idea a step further. David Rittenhouse was born in 1732 to a prominent Philadelphia family, and at the age of 17 he began building clocks and a variety of scientific instruments, performed early experiments with diffraction gratings, later used in stellar spectroscopy, and constructed a crank-operated replica of the solar system that sparked his interest in astronomy. After joining the American Philosophical Society, Rittenhouse was given command of a rare and exciting observing opportunity: the 1769 transit of Venus.

Transits were important in early astronomy. By observing the path of one of the interior planets across the sun, much information could be gleaned. Transits were useful in triangulating positions, especially the precise distance between the earth and the sun, and these events were eagerly anticipated. Three teams were formed. One would observe from a lighthouse on Delaware Bay. The second, using locally constructed telescopes, watched from outside the Philadelphia statehouse. The third team, led by Rittenhouse himself, did something that nobody else had ever done in America before.

They constructed an observatory.

They broke ground for the log building in Norriton Township in November of 1768, and it was completed, just in the nick of time for the event, in April of 1769. But by 1770, the observatory was torn down and the instruments moved to Rittenhouse's new observatory, a small brick building located on the corner of Arch and Seventh in Philadelphia, where he worked until his death in 1796. It would be almost forty years until another observatory replaced it.

There was interest, though. Congress passed an act on February 10, 1807, under the suggestion of President Thomas Jefferson, an amateur scientist in his own right, that a coastal survey be established. The survey would ascertain the true position of the coast by using astronomical observations made from a new and as yet unplanned facility: a national observatory.

The idea was kicked around during the terms of James Monroe and John Quincy Adams, the latter of whom on December 6, 1825, in his first annual message to Congress, remarked:

Connected with the establishment of a university, or separate from it, might be undertaken the erection of an astronomical observatory, with provision for the support of an astronomer, to be in constant attendance of observation upon the phenomena of the heavens; and for the periodical publication of his observations. It is with no feeling of pride, as an American, that the remark may be made that, on the comparatively small territorial surface of Europe, there are existing upward of one hundred and thirty of these lighthouses of the skies; while throughout the whole American hemisphere there is not one.

But over the next fifteen years, things would change. After losing a telescope in a shipwreck in 1822, Sheldon Clark, an early patron of Yale University, raised the money necessary to procure a new five-inch refractor from England. By 1830, the telescope was shipped and in the hands of students and professors in New Haven, who placed it in one of the tallest buildings on campus — the Athenaeum tower. The tower wasn't ideal. The telescope had to be wheeled from window to window on casters, which restricted its use for objects lower than 30 degrees above the horizon. But this didn't prevent astronomers at Yale from making the first American sighting of Halley's comet on its return orbit in 1835. A telescope, it was soon realized, needed a dedicated facility. It needed a building of its very own.

Joseph Caldwell, the first president of the University of North Carolina, came to his position with a dream: to purchase "an astronomical clock, a transit instrument and an astronomical telescope." He did exactly that. In 1824, with money wheedled from a reluctant board of trustees, Caldwell purchased a small telescope from London. It was hardly bigger than my Sears refractor.

But it was a start. In 1830, with funds from his own pocket, Caldwell housed the telescope and the accompanying transit instruments in an observatory — a tower and cabin built especially for this purpose. Unfortunately, the construction was shoddy, and a few years later the place was abandoned. It later burned to the ground. But an observatory was now in the public consciousness, and the next few years would see the construction of the first permanent facility, in Williamstown, Massachusetts, in 1838, at Williams College, which owned the largest telescope in America at the time — a ten-inch reflector. The observatory, built by Professor Albert Hopkins, still stands, the oldest facility in the country and one with a delightful history. On November 14, 1865, the writer Ralph Waldo Emerson said after visiting:

> Every fixture and instrument in the building, every nail and pin, has a direct reference to the Milky Way, the fixed stars, and the nebulae, and we leave Massachusetts and the Americas and history outside at the door when we come in.

By 1843, the Cincinnati observatory had a new 12-inch refractor located on Mount Ida, housed in a nice domed facility perched high above the city. The cornerstone was laid in cold and pouring rain by a now aging John Quincy Adams, who had traveled to Cincinnati in poor health, inspired by America's latest lighthouse to the skies.

Soon, observatory construction reached a feverish pace. Interest in astronomy boomed after the sighting of a comet in 1843, an object so brilliant in the March skies that it could be seen in broad daylight. "The greatest sensation," wrote noted author and abolitionist Moncure Daniel Conway in his autobiography, "was caused by the comet of 1843." William Miller, an itinerant preacher from Vermont and the founder of the Adventist movement, had prophesied that the world would end that year, and no doubt the comet seemed to bear him out. His followers, nearly fifty thousand by late summer, culled from tent shows and traveling revivals, climbed to the tops of roofs to be closer to the sky, and entire towns' populations were on edge.

"There was a widespread panic," wrote Conway. "Apprehending the approach of Judgment Day, crowds besieged the shop of Mr. Petty, our preaching tailor, invoking his prayers."

But America was expanding in the 1840s, reaching westward with the annexation of Texas, New Mexico, and California, and a confluence of political and scientific events helped to spark a popular interest in technology, a kind of miniature techno boom that culminated in Samuel Morse's first demonstration of the telegraph, between Baltimore and Washington, in 1843. It was a technology that would link a sprawling nation together. Technological progress began to balance the newspaper accounts of supernatural phenomena, trances, and mesmerism that inspired the literature of Poe, among others.

Observatories became an extension of that technology as many universities and colleges began to draw up plans for their own instruments and the places that would house them. Facilities at Harvard, West Point, Albany, and the U.S. Naval Observatory, among others, all began construction between 1835 and 1845, setting the pace and foundation for America's first and lasting love.

Science.

Perhaps John Q. Adams knew what every backyard astronomer has also discovered.

The night sky is a canopy of dreams. While at the eyepiece, the astronomer finds that his mind wanders and slips into the crack of imaginary time. Eons have passed since the light left most of the objects we observe, as eons will pass when yesteryears become tomorrow. Time is a shadow in space, fleeting, a wanderer who brokers in distances that few of us can comprehend. These are cartoon numbers, distant miles that are condensed into light-years and parsecs — numbers that can easily boggle the imagination. How can things possibly stretch that far?

That they do is a testament to the thirst of creation. To fill every spot, the universe careens and swoons like a carnival ride as it expands. I recall the words of a college professor when I asked him exactly what the universe was traveling into, as illustrated by the famous balloon analogy. You know it: a balloon, with painted dots representing galaxies, is filled with air, and as it expands, the spaces between the dots increase, a metaphor for universal expansion. But what does the balloon fill into? I asked.

With glee in his eyes, the cosmologist stretched out his arms.

"Into itself," he said. "The universe is expanding into itself."

It was the first time that I felt a confirmation of what my mother had always told me. The universe was a mystical place at best, an uncertain one at worst, and scientists were just discovering how weird and wonderful it all was.

Cosmology, since my childhood foray into astronomy, has increasingly overlapped the realm of science fiction. The ideas of time travel, extra dimensions, and parallel universes are quickly becoming the building blocks of the new cosmology. Many of the mathematical concepts of the universe — say, equaling the amount of lost matter that originated at the Big Bang — are introducing radical methods to make everything compute. Superstring theory alone demands new mathematical variables, and the cosmologists are opening their brains further and reaching deeper than ever before.

And when the professionals go somewhere, the amateurs are not far behind. CCD cameras, titanium-oxide–filtered photometers, computer simulations, and hydrogen-alpha solar filters were all once the protected technologies of the professional astronomer. No longer. The amateurs of the twenty-first century are better educated and better equipped than ever before. They are sitting in their backyards and taking photographs from small telescopes, with the help of computers, that would have rivaled the largest and finest instruments in the world just a generation ago. They are also relaxing under the domes and roll-off roofs of their computer-controlled observatories with probably one thought on their minds: what a great time this is to observe the heavens.

But I wasn't doing much of that. A cloud bank had settled in. Every year in Connecticut, the spring seems to usher in waves of rain and clouds that linger, then mysteriously withdraw. Most years I take this opportunity to seed my lawn — a near-worthless task, for the lawn is always dead by August, although this year I took a different tack. I dug a hole.

The ground was now level. I had towed, by force of muscle and wheelbarrow, a small mountain of soil from one part of the yard to another. This, I discovered, was to be the least of my labors.

To reduce vibrations at the eyepiece, telescope makers have gone to great lengths to construct better tripods. But even a tripod, as good as they now are, lacks the dampening qualities necessary for an astronomical observatory. Tap a telescope hard and you'll find a slight wobble, like Jell-O rolling

163

back and forth, an effect greatly reduced by a pier.

Most smaller observatories opt for a layer of concrete for their pier, in the form of either a submerged sonotube or a large slab separated from the observatory floor, to further minimize vibrations. From there a metal pier can be introduced, bolted atop the concrete and set below a deep frost line — my plan exactly. Barring one of William Miller's end-of-the-world scenarios, like an earthquake or meteor strike, the concrete shouldn't shift. It would also be heavy, lowering the telescope's center of gravity — if digging the hole didn't lower my own gravitational center first.

I pulled out my tools and started the long burrow. Easy. The first layer, perhaps a foot of car-packed soil used for leveling, crumbled under my spade. Then I hit. Stone.

The process of extraction, not unlike digging out a rotten tooth, consists of using a straight iron and a fulcrum to loosen the rock, then slow wiggling. Back and forth.

Curious, my daughters watched me. Every child and dog loves to dig, and every parent forbids it, unless at the beach, but I decided to splurge. I gave them a shovel and told them to dig.

"To China?" they asked.

Why not? With big grins, they went to work as I watched. I yanked out all the rocks myself, of course, as they carved a tedious circle around where the pier would sit.

For the next week I labored on the hole, uncovering rocks as an archaeologist would, with picks and trowels. I used pole diggers, pole punchers, poleaxes. The deeper I got, the worse it became. At three and a half feet down, I was in past my arm, wrestling with a boulder that was sleeping quietly at the bottom of my pit. And the pit was winning.

My youngest daughter came up with an idea.

"You hold me, Daddy. I'll grab it and you pull me up. Like this."

She illustrated by standing on her head, wiggling a foot skyward. I was to drop her in upside down, Houdini style, then lift her, presumably with the rock in her hands.

As a measure of my desperation, I actually considered this proposal. It was the logic of a child who watched too much television. Simply hold on to a foot and lower. They do it in cartoons. Why not in real life?

Demurring, I wrenched and contorted myself down the hole like a rabbit, or like Houdini again, in one of his escape boxes, just far enough in to be unable to work or grab anything as heavy as a stone the size of a watermelon. It was impossible. And I was only three inches away from making code.

But building codes must yield to common sense. What was three inches to a meager 80-pound telescope and pier?

I took one more measurement, and somehow those inches evaporated.

"We're there, kids," I exclaimed. "Four feet."

If the concrete heaves, then so be it.

Builders will tell you that for every job there is a tool. I owned the arcane utensils of the gravedigger's trade — spade, iron, and wheelbarrow. What I did not own were any practical tools.

My measly allotment of two hundred dollars for hardware was exhausted within five minutes of shopping. It was the man in the orange vest again, and I followed him through a small kiosk where they housed most of the power tools.

"You'll need a good drill. Extra batteries too."

"Can't I wait for it to charge?"

"You can, and sit around all day."

He was right. Two batteries. I picked out the cheapest drill the store sold, which was also on sale, and dropped it into my cart. A frown crossed the man's face, but he didn't say anything. It was obvious what kind of customer I was.

The cheap kind.

"Don't forget saws. Circular, miter. A handsaw wouldn't hurt."

Three more items. Then he gave me an L-square.

"For the rafters," he said.

Around this time I began to think about saving money.

"Do I really need all this?"

"They're angles. You'll do these how?"

Into the cart it went, along with a new hammer, drill bits, a plywood-saw blade, two tape measures ("You'll lose one"), a long level, a short level, a leveling bob, two packs of sandpaper, an assortment of goggles and earmuffs, and finally a tool belt. Then he stopped.

"I guess we're done," I said, hopefully.

But he just smiled. "You haven't bought any supplies yet."

The cart was overflowing. I thought these were supplies.

"You'll need screws. Plus nails. Tenpenny, twelvepenny —"

That was it. He was breaking me, and fast.

"I'm not spending twelve cents on a nail," I said.

Hardware stores are a lot like doctors' offices. Both deal with people at their absolute lowest. Several generations ago, when families were raised on farms, only the most inept man was unable to fix a combustion engine, repair a roof, or build a barn. Today, men have been separated from the land, and from the stars as well, working in cities and skyscrapers, and toiling behind keyboards — a job that was unimaginable for our grandfathers. Typing, once the secret domain of writers and secretaries, has become commonplace. Everyone can do it now: men, women, and children. We've advanced off the farm, though the knowledge and self-sufficiency gathered from farmwork hasn't followed us. We find it again at hardware stores.

"Twelvepenny. It's weight, not cost."

Nails, he said, have a fascinating history. They have been around since the ancient Egyptians and have long been prized. Before the common throwaways used at modern building sites, the utility and value of nails were recognized by our revolutionary ancestors, who had to purchase them in Britain before Yankee ingenuity took over. Working at fireplaces and forges all across the thirteen colonies were nail-makers — amateur mom-and-pop smelting operations that drew in some of the elite of

American society, including Thomas Jefferson, who, in 1794, started a nailery at his home at Monticello.

The measure of weight for nails — the penny — was derived from English currency. A pile of one hundred 3¹/₂-inch nails might cost around sixteen pence — or sixteen pennies. The name stuck, though sometimes it's replaced by the letter *d,* for denarius, a small Roman coin. Obviously I wasn't the only person who thought nails were expensive.

"They increase by quarter-inches," he said, "from two-d to sixteen-d. Use these to anchor. Don't forget screws. Galvanized, stainless steel. These won't rust."

He pulled out six boxes and stuffed them into my cart.

"This should get you started."

"Just started?"

"Oh, you'll be back. They always come back."

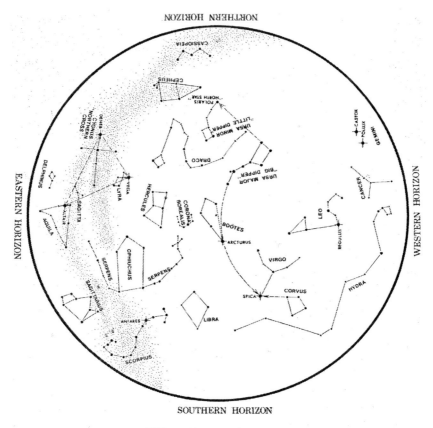

THE NIGHT SKY IN JUNE

June

The first star tonight is Arcturus, in Boötes — or is it Vega, in Lyra? My children, who are looking in opposite directions, as lost sailors do — one to the west, toward sunset, and the other to the east, toward night — argue.

"I saw it first," says the oldest. "The blue one."

Vega. It's dimmer than Arcturus by a smidgen.

The youngest holds her ground.

"But you can see the red star better."

"Mine was first."

"Only because your sky is dark."

I try to referee. Both stars are bright. They creep out at sunset, as the world shifts to halftones, silhouettes, varying shades of black. It's the coming of night, and soon the argument is over. More stars show up, distant beacons, first two, then six, now a congestion.

Stretching from Arcturus, and linking both summer and spring sky, is the constellation of Boötes. Boötes is the boy in the field, the shepherd, who travels far from his village. Between the constellations of Leo and Boötes lies a great vastness, an empty hole almost, devoid of bright stars. But Boötes is easily seen. With its large kite-shaped appearance, Boötes stretches up in the June sky, near the zenith at dusk, and calls for his sheep. It is night, and he is alone, and all the sheep have wandered off, audible as a faint bleating over the hill.

The boy whistles. From the darkness stirs one of his dogs, Canes Venatici, a scrawny mutt who runs from the fields, as dogs often do, smelling something. Bear. Situated between Ursa Major and Boötes, Canes Venatici is one of the fainter constellations in the sky, noticeable only by the bright double star Cor Caroli, though there are other wonderful objects. M3, the great globular cluster, is located just northwest of Arcturus and is easily visible through binoculars. I imagine M3 as sugar spilled at breakfast and heaped at the table by tiny fingers trying to disguise an accident.

On the other side of Canes Venatici, just south of Alkaid, in the dipper handle, is M51, the Whirlpool galaxy. Its spiral arms face us head-on not as one blur of light but two. M51 is a car crash, a wreck on the interstate. A second galaxy, near the spiral edge, has run a red light and now is hanging from the side of the road, a junker in need of a tow. Two galaxies have collided.

Fortunately, help is close by. Corona Borealis, at first blush, resembles the lowering hook from a tow truck, with its clipped semicircle of bright stars on the opposite side of Boötes, but that image would be misleading. Corona Borealis is the Northern Crown. The Mercedes of constellations, the Crown is a wondrous selection of double stars, like Sigma Coronae Borealis, and variables. These are stars that change in brightness for a variety of reasons: an eclipsing companion or the depletion of an atomic core that causes the star to heave off layers, dimming and surging as flashlights do in the cold.

One of these variables, R Coronae Borealis, is in a separate category unto itself, a kind of anti-nova surrounded by a deposit of soot. Most novas, or stars that violently change magnitudes, tend to brighten, but not this one. R Coronae stays constant for months or even years on end, at nearly naked-eye levels, and then it mysteriously falls off the cliff, becoming so faint that it takes a good telescope to ferret it out.

There are many such secrets in the

Northern Crown. Sweep through the area with a pair of binoculars and a few are suggestive. Stars that suddenly split, or even stars that wish to move. The reason? Man-made satellites in Earth's orbit. They were around in the '70s, of course, but there were nowhere near as many as today. I shouldn't be surprised. Global positioning systems, weather satellites — communication, defense, research: Earth is surrounded by flying metal, and it all seems to cross one vector more than any other: right through Corona Borealis.

While keeping tabs on my variables, I notice wave after wave of moving objects in my telescope. Hardly UFOs, these are neatly scripted sentinels, artificial stars on a monotonous course. But they aren't the only intruders from space. There is another one, brighter than any visible satellite, its slow trail resembling the falling spark from a Roman candle.

"Is that it?"

My youngest daughter is pointing again, but not at Arcturus. I check my watch, shake my head. Too soon. She's a first-grader, interested in science, and we're outside angling to catch a glimpse of the International Space Station as it passes overhead, but mostly we're only catching the interest of mosquitoes.

"Can we see the astronauts from here?" she asks.

"No. They're too far."

"But someone's driving, right?"

"Gravity is. Imagine a racetrack, round and round."

"Sounds boring," she quips.

It does, even though it shouldn't. My first space station was Skylab in 1975, and I recall climbing the stairs of a local water tower to reach the top. This wasn't legal, of course. People were arrested all the time, but the tower was the tallest structure I could find — perfect for a Skylab flyover.

This wasn't my first climb. A year earlier, I had spent an unsuccessful week looking for Comet Kohoutek, but my confidence in finding Skylab that night was high. I had charts of orbital paths, and I recall sweeping the horizon, unsure of what I was looking for. But as a teenager and amateur astronomer, I knew one thing. This was a first, a real American space station; fancier than the Soviet Soyuz, it was right out of the comics and science-fiction novels that I loved. My excitement built as I waited, and then came the disappointment. I never found it.

"What will it look like, Daddy?"

This time I had an answer.

"Probably like a star, only moving."

She points at something. I check my calculations again. Wrong direction.

"Do the astronauts ever see aliens?"

"I doubt it."

"But there are aliens, right?"

I shrug and tell her that the astronauts are busy. There are experiments to be done that preclude the search for aliens. But my daughter doesn't buy it. She's watched television. *Star Trek. The Twilight Zone.* What was more important than invaders from space?

"If there are aliens," she says, "they'll find them. Can we see them in the telescope?"

I'm confused. "The aliens?"

"No. The space station."

"I don't know. It's pretty fast."

"Like a rocket?" She zips her hand across the sky, accompanying it with a whistling sound that only kids can make.

"Slower, I think. I've never seen it."

"Never?"

How was that possible? Hadn't I spent the last eight months outside?

I'm about to answer, to tell her that there are many things that I haven't seen in my years of observing, including Skylab, when suddenly we notice it — the International Space Station. It comes in low, just above the trees, a bright, starlike object moving steadier than any surgeon's knife.

"That's it?" Her disappointment echoes.

"Imagine. The largest space station ever built."

The ISS is a modern miracle, I tell her, now one of the brightest objects in the heavens, a fact made even more incredible by this: the ISS was made by human beings.

"But it just looks like a dumb plane."

"Now it does, sure, but not in space."

She swats a few mosquitoes. The imagination of a child is vast, uncharted territory where no adult should tread. The idea of a space station conjures in her mind gigantic cities, a sprawling metropolis of light that can easily be seen, instead of the prosaic reality. Something that looks like a plane.

Technology impacts a generation with memory of it least, like my father shaking his head at the miracle of computers while I smile and nod expectantly. My memory of a world without technology isn't as long as his. This is a problem that my daughter won't have. Thanks to the ISS, orbiting space stations are now as much a part of her life as pizza, video games, and digital pets, and sometimes just as mundane.

"Do they eat in space?" she asks.

"Certainly."

"Good. I'm hungry. Can we go?"

It's been said that telescopes are instruments as much of disappointment as they are of awe. The sky is rarely settled for a good view, especially for an impatient beginner, and the imagination of a new stargazer almost always outstrips reality. Objects in a telescope are smaller than expected, not as clear as in a photograph, and always too dim. Nowhere is this more apparent than in a continued journey through the spring sky.

Compared to the clean geometry of the Northern Crown, the constellation west of Boötes is terribly faint. This is Coma Berenices — Berenice's hair. A waterfall of stars, Coma Berenices has been called an asterism, or grouping of stars, rather than an actual constellation. For several thousand years, the stars in Coma Berenices have bounced under the ownership of Leo to Virgo in the eyes of celestial cartographers. By itself, Coma Berenices is cosmic filler, a bureaucratic confusion of stars who, without sharing part of the Virgo-Coma cluster, might easily slip into oblivion.

But beauty is timeless, and the long mane of Coma Berenices seduces even the casual stargazer. It's home to a famous binocular star cluster, Melotte 111 — a loose assemblage of about eighty stars that riddle the eyepiece like bullets in a gangster's windshield. This is Berenice's hair. According to legend, a queen offered up to Aphrodite her long tresses for the safe return of her husband from battle. But the hair vanished from the temple, and the court astronomer, Conon, angling perhaps to keep his job, claimed that the queen's hair had been transformed magically into a constellation by the gods. On the strength of stolen locks, then, the work of ancient fans, perhaps, or paparazzi, a story was born.

The story is a common one. A boy, Boötes, grows up in a small town and takes the only job that he can find, as a herder. But he isn't happy. A girl has just passed by on her way to the city, and she is the most ravishing creature he's ever seen. Soon there are temptations that every herdsman experiences, especially a young one — the call to leave his land. Money can be made in the city. Golden crowns bought, beautiful women courted. But what about his sheep? There are dangers in the fields. A serpent, the dreaded constellation Serpens, coils its

179

way towards the unsuspecting sleep of Boötes as he dreams of a better life.

A weaving, slithering line of stars to the south of Boötes, the serpent keeps his pointed head low to the ground. All stealth, Serpens creeps up on the stargazer, often late on a June night, when the sky is darker and his eyes are better adapted. Then they pop out, the keen pupils of a reptile.

At the neck shines the brightest star of Serpens, called Unukalhai by the Arabs, and I see it as a warning star. The snake has a ring around his neck, no doubt a deadly species, and he's wandering in the grass alone. Serpens is a split constellation. He is cut in half by a tent-shaped figure, Ophiuchus. Cloaked in mystery, Ophiuchus holds the snake, commands it to do its deadly task. He is the snake charmer, the releaser of poisons, and Boötes must fear for his sheep and for his life.

But in the west, growing higher off the horizon on this June evening, there is a light. Like a campfire, Venus throws cheer into the sky. No planet is brighter than Venus, nor as mysterious. Venus rarely leaves the comfort and security of the horizon. At her highest point, seen from northern latitudes like mine, forty degrees is about tops — about the level of tall trees from a short distance away. But Venus, unlike many other targets, can never be mistaken. She is the third-brightest object in the entire sky, behind only the sun and the moon. She has been known to cast shadows on the earth, to inspire fear in the ancients.

Obviously the ancients never looked at Venus through a telescope or they surely would have been disappointed. The brightest planet in our solar system shows phases exactly like the moon's but there the similarity ends. Venus is choked by clouds, and surface detail is nearly impossible to see.

The appearance of Venus is brilliant: a dazzling diamond white that in most telescopes looks as bleached as Georgia O'Keeffe's paintings of skulls. Over the years, enthusiastic stargazers have reported seeing faint shadows, grades of brightness, and phase deformations, despite all the clouds. The eyes of an amateur astronomer are his most valuable instrument and one that is often the most difficult to measure objectively. When compared to those of another backyard enthusiast, the task is far easier, though no two sets of eyes are the same, and some astronomers can simply see better than most of us. But observing Venus is like looking at a fun-house mirror.

She has fooled people before, both famous and obscure.

Take Johann Schröter.

Schröter's career has often been compared, if unfavorably, to that of William Herschel. The two men were contemporaries, and both were pioneers in observational astronomy although trained in other disciplines, and both were German. Herschel had moved to England from Hanover in 1757 to pursue a career in music, but he soon developed an interest in astronomy that culminated in the discovery of Uranus in 1781, which made Herschel one of the most famous astronomers in Europe. Schröter was not. His own path into history proved difficult, slowed by famous students and an eccentric temperament.

Schröter was born in 1745 and educated at the University of Göttingen. He studied law, and was soon employed at a legal firm in Hanover, where he met Herschel's youngest brother, Dietrich, who had just returned from Britain with incredible news. His brother, William, had recently built a telescope and was busy observing the heavens.

Perhaps inspired by William Herschel, or simply curious, Schröter procured his first telescope in 1779, a three-foot Dollond refractor that he used to observe the sun and moon. Four years later, he accepted an appointment that would change his life and the lives of his future students. He became chief magistrate of the town of Lilienthel, and there he undertook a new project: the building of his own private observatory.

His work at the observatory was erratic, prone to wild changes of direction. A fierce proponent of the existence of extraterrestrial life, Schröter took observations of light changes on the moon to be signs of colonization, potential cities. "I at least imagine," he wrote, the "gray surface of Mare Imbrium to be just as fruitful as the Campanian plain." He believed that Venus was inhabited and that the markings on Mars were largely impermanent.

He also saw mountains where there were none.

From December of 1789 to January of 1790, Schröter noticed a separation of the southern horn of Venus that looked like a bright detached island, and he mistook this for a mountain that he estimated to be some twenty-seven miles high. This was a view ridiculed by none other than Herschel himself.

"As to the mountains of Venus," he wrote, "I may venture to say that no eye which is not considerably better than mine, or assisted by better instruments, will ever get a sight of them."

Since Herschel was regarded as having the best eye and the best instruments in the world, his deduction was obvious: the mountains didn't exist.

Such criticism would have haunted some men, but not Schröter. His passion for observing was dogged and even inspirational, especially to the busy modern-day amateur. Schröter worked long days as a lawyer, then spent his evenings at the eyepiece with few of the rewards that were bestowed upon Herschel — except for a moniker that surely would have rankled the discoverer of Uranus. The Herschel of Germany.

Soon he wouldn't even be that. On the evening of April 20, 1813, invading French forces ransacked Schröter's home, burning his papers and books and destroying the observatory. Schröter never recovered from the loss, and he died a few years later.

Three decades of observational work went up in flames that night in 1813, including all of the astronomer's notebooks, dating back to 1779. Some recorded the first map of Mercury, others the differences of atmospheric rotation on Jupiter. Exactly what else was in those notebooks nobody knows. Erratic musings, or wonderful discoveries? Perhaps some of each, for great men, even in the full glare of their errors, are still subject to the natural laws that all of us face. They are still human and still fallible. They are always immensely interesting.

Spring ends with a warm blast.

In the first week of June, temperatures soar to unseasonable heights as I receive my first load of lumber for the observatory. The ground has been leveled and the pier dug as I pound my first nail — for the floor supports — in the sticky dusk.

Foundations are important, not just in the lives of our children, but in construction. An uneven foundation invites disaster. A rotten house can be saved, given a solid foundation, but a good house over a rotten foundation is doomed. Foundations always win. And, like the averages of baseball players and stock pickers, they are irrefutable. A foundation either does the job or it doesn't.

So it was that I spent the extra energy making sure that the runners, great six-inch-by-six-inch beams that would eventually support the floor, were actually level. These were pressure-treated timbers, now resting on a neat layer of pea gravel that I lugged in, bag by perilous bag, as a precaution suggested by my brother-in-law.

"You don't want a shifting floor, do you?"

I had dirt. Doesn't that count?

"No. Dirt can wash away. You want a stone base."

Try telling my aching back it didn't count. But I listened to him anyway and purchased several carloads of ground stone. Except for the weight involved, smoothing out the gravel was easy. Simply sweep and rake, as a monk at a Kyoto rock garden would do, then step away. No sweat, I thought, until I set down the first two beams.

They weren't level.

"Measure from the corners," said Jim. "Align those first."

"How exact should I be?"

"Like you're building the space shuttle."

The unpacked gravel under the beams had created deep, periodontal pockets, holes that allowed my hands to go under the wood and into loose rock. In other words, these beams were resting on a hope and a prayer.

"You need even contact," he said, "or the floor will lean. The roof too."

Gravel, it seems, adds a peculiar mix to a building project. Compacted soil is stationary; it doesn't give way. But gravel is fluid, like a sand dune; it rolls and slowly lists in any direction, as I quickly found out. Gravel is also malleable; it can be pushed into holes, worked as filler, and leveled, all with the cadence of a midnight road crew.

For the past week, noise has been coming from our house, early in the morning before work, and again at twilight. Unusual sounds. There were rumors: a massive addition done on the cheap, an enormous sandbox for my girls, a doghouse. The talk drew the curiosity of a few neighbors, including one friend, a bond trader, whose lands abuts mine. He asked me what I was doing.

"I'm building an observatory," I said, "for looking at the stars."

A big grin crossed his face, slowly, like a solar transit.

"What are you *really* doing?"

He glanced around and saw what looked like a lumberyard. A five-foot-high stack of wood, too much for a doghouse, not enough for a family room.

"You're kidding me, right?"

I told him that I wasn't.

"I'm going to put a telescope in here, maybe a CCD camera."

He nodded and walked in a circle around the foundation, considering my words like the footnotes of an annual report, analyzing them.

"You can't see anything. The trees."

I shrugged. There was always a chainsaw.

"Then there's New York. Isn't the sky too bright?"

"It's not always bright."

He could go on. Rain, bugs, snow. And he probably did, in his own mind, before leaving me with one of his classic smiles.

"Oh, I get it. You're building a boat in the desert."

He nodded and walked away, laughing. The project was ridiculous, but maybe it was an inspiration as well, for a year later my friend would level a third of an acre himself to plant something that he'd always dreamed of, from his childhood in Japan: a bamboo forest.

Whimsy is infectious. It originates in the games of children and fans out, first with pretend tea parties and playing with trucks, then with imaginary playmates, painting, all the arts. Writers understand whimsy. Our work is built upon conversations with ourselves, which isn't very practical. Every day we conjure up characters, cultivate faint possibilities, juggle caution.

But writers and artists have staked no claims to whimsy. The ridiculous is everywhere. At Halloween, in my town, there are people who go overboard by building theater sets in their yards, fake pirate ships complete with gallows, a plank, the flapping Jolly Roger. The desire for meaningless play is a strong one. All across America, whimsy rules. There are corn palaces in South Dakota. Giant toy-railroad farms in Pennsylvania. Statues of Paul Bunyan in Minnesota. The lure of impracticality connects with our sense of humor, and for this purpose: it has none.

Science isn't immune. Of all the hard sciences, astronomy is perhaps the most unnecessary, and like fiction, it is a pursuit that appears to have little impact upon everyday life. Why should anyone care how old the universe is? What possible bearing does understanding the light curve of an eclipsing binary star have on earthly chores? These are tough questions. But science is more than just a rat's scurry for knowledge; it also helps us to illuminate ourselves as human beings.

Astronomers embrace the impossible and celebrate the patently absurd. We observe at night when most of the world is sleeping, and the battles that we wage are against nature — clouds, dew, an unsteady atmosphere. Sometimes just the effort required to leave a warm bed is Herculean, demanding a kind of zaniness that extends into an astronomer's projects.

In 1845, Lord Rosse, a stargazer and the wealthy landowner of Birr Castle, Ireland, decided to build the largest telescope in the world. It was a giant reflector that needed a small army to operate it, and then only with great effort. But for two decades the 72-inch telescope contributed greatly to astronomy, and it reigned supreme despite its cost and difficulty to construct.

Much earlier, the great William Herschel tried his hand at the largest-telescope game. He built his monstrosity in his yard as well: a lattice of timber and glass, the telescope was several stories high, stopping all traffic. People got out of their horse-drawn carriages. Young children climbed over fences. Women gossiped. Everyone must have been telepathic. What is this nut doing?

Easy. He was building a boat in the desert.

The dream of building the largest telescope in the world, not unlike that of building the highest skyscraper or the biggest ship, has been with us for a long time.

The spidery aerial telescopes of the seventeenth century constructed by Johannes Hevelius soon revealed a serious design flaw. Refractors could only be built to a certain length because the glass required for their objective lenses had to be cut in a way that limited their size. The answer was mirrors.

In 1789, William Herschel decided to grind and build the largest speculum mirror ever — 48 inches in diameter. The telescope would be the culmination of his career; it would also be a disaster. On the rare occasions when it could be used, wind and rapid changes in temperature affected the image quality. But more worrisome was the size.

Caroline Herschel, William's sister and an astronomer in her own right, wrote in her diary on September 22, 1806, "In taking the 40 foot mirror out of the tube, the beam in which the tackle is fixed broke in the middle. . . . Both of my brothers had a narrow escape of being crushed to death."

The great telescope was never used after 1811, but the lure of huge telescopes remained. Williams Parsons, the third Earl of Rosse, designed and built an instrument that was used to discover the spiral nature of galaxies. This was the great "Leviathan of Parsonstown" — a four-ton, 72-inch telescope mirror and tube, anchored between two stone walls and hoisted into place by a complicated system of ropes and pulleys.

Between 1845 and 1917, the Rosse telescope was arguably the largest in the world, a title it held longer than any other telescope before or since. This fact would no doubt have tickled the third earl, who, perhaps ignominiously, ushered in a new trend in observatory construction and design: it became exclusively for the wealthy.

Wealth often carries with it a few eccentricities. Consider James Lick, the richest man in gold-rush California, who in 1874 directed in his will the building of "a telescope superior to and more powerful than any telescope yet made . . . and also a suitable observatory connected therewith."

Born in 1796, Lick was a man touched by fortune. He was a master craftsman, a skilled woodworker who built pianos and organs in New York City only to learn that these were being shipped to South America, where there was a great demand. So he moved to Argentina. There, surrounded by political turmoil, Lick found both a company and trouble of his own. On a return voyage home from Europe, he was captured by a Portuguese man-o'-war. The crew and passengers were taken to Montevideo as prisoners of war, but Lick escaped.

Perhaps it was this close shave with death that inspired him to come to San Francisco, or maybe he sensed his future. On arrival in 1848, Lick began to buy large blocks of the city at prices that would skyrocket only seventeen days later when gold was found at Sutter's Mill. The gold rush was on, and James Lick stood to profit: he supplied real estate for hotels, boardinghouses, and saloons to a town that grew from a population of little more than a thousand to more than twenty thousand inhabitants in less than two years.

But by 1870, he was an old man and looking to disperse his wealth. An astronomer friend of Lick's, George Davidson, reportedly came up with the idea to secure the old man's posterity. A telescope. Lick wasn't sure. He had several options on the table. One was a giant statue of himself and his parents, not unlike the famed Colossus of Rhodes, that could be seen overlooking San Francisco Bay by passing sailors and merchant ships. Another idea was a pyramid, larger even than Cheops, that Lick would construct downtown. Thankfully, neither of these made the cut, and James Lick settled for a telescope, which, in the final irony of his life, carried a caveat — that it be the largest working refractor in the world.

Soon it wouldn't even be that.

In the late 1890s, a new telescope was

conceived, the brainchild of one of the most influential men ever to come into astronomy. His name was George Ellery Hale.

Hale was born in Chicago in 1868, and he became fascinated with astronomy around the age of 13. "I learned of a second hand Clark refractor," he wrote, "of 4 inches aperture. This was purchased by my father and mounted on the roof of the house. . . . Thus I became an amateur astronomer."

Like many amateur astronomers, Hale loved not only the science of astronomy but also the instruments. No sooner had he mastered one telescope than he began to dream about the next, always larger. Eventually this led to a small observatory of his own, where he studied the sun, his continuing passion. He invented the world's first spectroheliograph, an instrument useful for photographing solar prominences, and he cofounded a magazine for professional astronomers called *Astrophysical Journal.* But George Hale wanted more.

At the age of 24, he met William Harper, the president of the new University of Chicago, who was going around hiring the brightest talent in the astronomical community. He came to Hale with an intriguing offer: would he like to work at the college, contingent upon the building of a new observatory? It was an offer and a path that would change Hale's life forever.

The two men joined forces and set their sights on a Chicago businessman, Charles Yerkes, who was looking for a way back into proper society. Twenty-five years earlier, Yerkes had been sent to prison for misappropriating funds from the city of Philadelphia, and he spent thirty-three months behind bars. But now Yerkes had rebuilt his business and was as prosperous as ever. Hale proposed to him an idea: the largest telescope in the world, financed by Yerkes and bearing his name.

What Hale was offering was a slice of immortality.

The Yerkes observatory was dedicated in 1897, and it remains the largest refracting telescope in the world. But refractors were already going the way of the dinosaur. Gigantic reflectors were in; these large mirrors could easily dwarf the light-gathering capacity of smaller refractors. Again, Hale was at the forefront. He moved to California and undertook an entirely new project at Mount Wilson, a 60-inch reflector that first saw light in 1908. No sooner was the Mount Wilson observatory finished than Hale concocted his new scheme. Financed by Los Angeles businessman John D. Hooker, the construction of the 100-inch Hooker telescope would surpass anything ever built. Although this instrument was the world's largest for almost three decades, it would prove to be a test run for Hale's crowning achievement — the great Mount Palomar observatory.

Palomar is a technological marvel. A delicate balance of brawn, brains, and striking good looks, this instrument has performed flawlessly for more than half of a century and continues to do so to this day, rewriting astronomical history. But Palomar is no longer the largest telescope in the world. Like heavyweights vying for the crown, there are always contenders. Now the champion reigns in Hawaii, at the Keck, and soon with the European VLT Interferometer at Cerro Paranal in the Chilean Andes, but others will come. They will arrive on land, or more likely, from space, as orbiting platforms. The Hubble telescope has already redefined the concept of the observatory, which has evolved much from the days of Stonehenge and Samarkand and will continue to evolve. But one thing will never change for the observatory. There will always be night, and the sky will always glow with countless radiant suns.

Summer

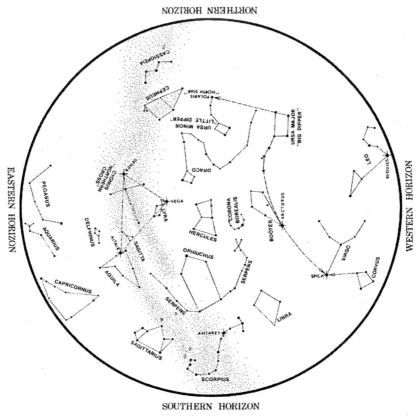

THE NIGHT SKY IN JULY

July

Tonight there are battles in the sky. Great warriors have risen in the east, and they are now at the zenith at the close of day, wielding clubs and dangerous arrows. One of the largest of these summer constellations, Hercules, stretches overhead. His body is taut and muscular, a block of stars surrounded by the pinwheeling arms and legs of a man too busy to sit. Hercules is all about motion. Located between Corona Borealis and Vega, Hercules, it seems, needs to move, like a film stuntman. Danger lurks in the heavens, and not just the danger that Boötes, the herder, had witnessed with the serpent. There are other, more formidable perils: monsters within.

Hercules is the hero, a knight, the keeper of the code. From his square body to his long L-shaped legs, Hercules lords over the July skies with the presence of a weary combat veteran. He knows that the quiet is only a lull in the fighting.

The herd of sheep is threatened in the

west, not just by Serpens, whose evil power was unleashed by Ophiuchus, but by another menace lurking in the weeds — Scorpio, the scorpion. A long fishhook of a constellation near the southern horizon, Scorpio has one eye, Antares, a red supergiant. Its color rivals Mars at opposition, and it gives the scorpion an ominous countenance. This is more than just an ordinary arachnid.

Many cultures, and their astrologers, have imbued Scorpio with mystical qualities. He was the Scorpion Man in Phoenicia, the Scorpion King in Egypt, a dragon in China. For the Toltecs this was Quetzalcoatl, the plumed serpent, and a force to be reckoned with. Bad news for Hercules. His twelve labors, penance for the accidental murder of his own family, have weakened him, leaving him vulnerable to human temptations. There's the love of the maiden Virgo, riches of the Northern Crown. And perhaps for the first time, Hercules ponders his situation. Will the fighting ever end? It's a question that all warriors ask themselves sooner or later, in the dark of night, with the enemy over the hill.

Through a telescope, Hercules offers the backyard stargazer one of the finest treats of the northern sky: M13, the great globular cluster. Summer is the time of the globulars. Rich condensations of stars wound in tight balls, globular clusters resemble an artist's dropcloth with paint everywhere. A five-inch telescope will show the glory of M13 almost down to her core. Stars peek out at the fringes, individual swaths of blue and yellow suns, as the eyepiece fills with a kaleidoscope of stars, thin tentacles and lanes that propeller from the center. In a larger telescope, the effect is almost three-dimensional. Eyes fall into it. The mind staggers to crawl out, even if it doesn't want to.

Finding M13 is easy. Binoculars will do it. First, locate the body of Hercules overhead, four bright stars in a trapezoid, and then continue south from the top star of the square about two degrees, or a finger's length away. In binoculars, M13 will look like an unfocused star, only larger, especially when compared to the other great globular in Hercules, M92.

Located ten degrees north of its larger, cosmopolitan cousin, M92 finds majesty in size. This is one of my favorite deep-sky objects. Its core appears to be tighter than that of M13, like that of a golf ball with its casing removed, and just as dangerous. There is a sense that M92 is combustible, a hand grenade of light that gravity can barely contain;

its pin has been removed and the seconds are counting down.

Move south and the conditions aren't much safer. Between Hercules and Scorpio lies a bounty of globular clusters, each one a menace. There are M10 and M12, in Ophiuchus. Ophiuchus resembles a hooded bank robber, or maybe he is cloaked in darkness, as movie villains sometimes are, walking through empty streets at night. Night rules Ophiuchus. His head is tucked and low, at the point of the constellation, and his arms are nonexistent, stuffed in a coat that hides both clusters, which are located near his heart, and also near to the heart of our galaxy.

When we look south on a summer evening, low towards Scorpio and Sagittarius next door, we are really gazing straight into the center of our own galaxy. Look at a photograph. The spiral arms of the Milky Way extend north into Cygnus and Cassiopeia, but the brightest part is to the south. This is the galactic center. Unfortunately, no view is perhaps more maligned for the amateur astronomer than the southern horizon. My south resembles melting ice cream, with the faint whitewash from the lights of Broadway, and below that, all trees.

I'm probably not alone. Darkness has

been replaced with light — electric, battery, fluorescent. Few of us have ever seen a truly dark sky before, far from towns and highways, so our point of reference has changed. We are all living on a grading curve, so impoverished by what we have seen that we simply can't imagine what we cannot see — the full majesty of the heavens. But it's there. The glory of the sky can be recaptured at any moment. Simply turn out the lights, unroll a blanket, and look up.

I'm looking up too. For the first time there is shape to my observatory. The floor is done, built and raised on its platform of gravel, and boxed in the center for my pier. I originally planned to put a refractor in the observatory, but a telephone call changed that. It came from a high school friend, Craig, who told me that he had bought a telescope, his first in twenty years.

Only months earlier, I'd e-mailed him the news. I was back into astronomy and observing as well, and he found this humorous. As teenagers we spent late evenings in his backyard, a dark site located just across the road from my future wife's home, so maybe my mother was partially correct — the influences of the stars do impact our future. What amazed me about Craig wasn't what he knew about astronomy, which was formidable; it was what he owned: a six-inch Criterion reflector. He was the only kid I knew who had a telescope.

Like me, Craig had left astronomy long ago, after college. He'd found a career in physics, and when I'd last seen him, at a book signing of mine, he'd given me the sad news. The Criterion was gone. Sold.

"The new telescope," he said, "is a large refractor."

He told me that the instrument was a difficult one.

"I observe sitting on my butt. The eyepiece is that low."

This was important. My pier would have to correspond to the height of the telescope, higher for a refractor, whose tubes were heavier and longer than those of many telescopes. Inexpensive refractors also show false color. Dense glows of purple that form around the perimeters of stars, planets, and the moon that make color photography troublesome, a problem that Craig's older telescope didn't have. Now a classic, these telescopes from the '70s were as coveted as restored GTOs, Schwinn Orange Crate bicycles, and original Barbies in the box.

Our past recirculates in the present. Sometimes the steps that we take are only shadows of ones already taken. I know. Only three months earlier I had ordered another telescope, a handsome, wide-field instrument for panoramic views of the Milky Way — a fact that stunned my friend.

"Another telescope?" I could hear his thoughts, grinding away like a bad declination motor. Not one telescope, but two, in six months. "You're set, then."

But I wasn't, and he could tell.

"I simply need more firepower," I said. "You know how it is."

Then I announced my new plan, one created on the fly, really, after hearing his refractor woes. If I had a larger telescope, like a Schmidt-Cassegrain, with its folded optics and squat design, I could hook up the electronics to my computer, an expensive option that now seemed attractive. With this telescope I could go deeper, gather more precious photons.

"You're nuts," he said.

And maybe I was. The observatory that I was building was surely indicative that I had caught the bug, and bad. It wasn't unusual. In my months back into the hobby I'd heard similar stories of people leaving astronomy for years and then returning, like a hurricane in opposition to the sea, sweeping in with greater punch. The reasons are as varied as people are: family responsibilities, illness, career pressures, divorce. But the stars are forgiving, and they are always there.

My yard, however, was less forgiving. It was changing beyond recognition. Trees have been trimmed, hacked, and culled. A small platform rose next to the house where before there was nothing but dead grass. I was used to that dead grass, and so was my wife. She inspected the floor, even stepping up on the risers.

"Is this going to be large enough?"

Like gravity, my project was slowly drawing her in. I pulled out a chair and pointed out the dimensions, including where the telescope would go, the new one that I hadn't told her about. Eight feet square. It was large enough.

"But the walls —"

One was up already.

"— they aren't very high."

A measuring tape snakes from my hip. High enough.

"Won't the telescope need to come up through the roof?"

A common misconception. Our mental picture of observatories are of the great instruments, especially the giant refractors, with a long tube emerging triumphant from the opening slit of a dome. It's an illusion. All telescopes, I told her, remain within the gentle bosom of observatory walls, which assist in keeping out a variety of menaces, both seen and unseen: dew, wind, animals, extraneous light. I hoped mine would do the same.

"I'm just saying," she said, "it can be taller. You're tall."

Her dress shoes tapped a circle around the hole.

"What's this?"

"For the pier. Imagine a pipe bolted in concrete."

She nodded. "Which telescope goes on that?"

I heard myself trying to answer, as though from a distance, like a ventriloquist practicing. The first telescope that I got, my wife had enthusiastically endorsed. It was a late Christmas present from her, and she gave me her blessings. But the second telescope, ordered only three months later, wasn't as well received.

"You already have a telescope," she'd said back then.

These were the same words my father had used. My answer was the same.

"But I need it. Photography."

"We aren't getting obsessive, are we?"

A book collector in California had died recently. When the authorities came in to settle his estate, they noticed that there were so many books in his house, stacked in every room clear to the ceiling, that the collector had actually been forced from his bed. He had been sleeping in a small corner of his unheated garage, a space that got so cold that he caught pneumonia and died. Did she mean that obsessed?

The second telescope passed muster by only the thinnest of margins. But even love won't save me on a third, I thought. Telescopes are like drinks at closing time. It's easy to go overboard late in the evening when you have no direction other than that from the compass of your own obsessions.

She smiles. These patterns of mine are so threadbare, too obvious not to see. "You're thinking of another telescope, aren't you?"

Astronomers' wives, I'm convinced, have the worst of all possible worlds. Unlike bowling or evening poker games, the night doesn't end. Even in the daytime, the lure of the stars persists in the glossy, four-color advertisements in *Sky & Telescope* magazine. The seduction is purely an American one, the promise of clever innovation at decreasing prices. But there is a hidden cost.

Some amateurs leave astronomy for reasons other than the few that I've mentioned. They simply burn out.

No one has expressed this feeling better than John Herschel, William's son and the reluctant heir apparent to the career of his father, the greatest observational astronomer ever.

"Two stars last night," wrote John Herschel in his diary in 1830, "and sat up till two waiting further. Ditto the night before. Sick of stargazing — mean to break the telescopes and melt the mirror."

Often a stargazer surrenders to the marketplace. Big eyepieces, spiffy new instruments, and cameras can easily exhaust an observer's wallet and his spirit. A beginner may feel that his equipment is inferior and will give up or, worse, believe that he has nothing to add to science without going into debt. Not true. The history of amateur astronomy is one of making do. At every star party in America there are creative hobbyists who push the limits of their equipment, and often do it for next to nothing.

I've known backyard astronomers who have built excellent telescopes from PVC tubes, dew heaters from stereo jacks, mounts from steam-fitting pipes. The desire to observe the heavens with materials made by hand is still a strong one. Expensive isn't always better. A stargazer must train every part of his body — his eyes, his mind, and his judgment — all of which are free.

Not that temptations don't lurk. Like the man who squeezes in one last book hoping to complete a collection that can never be completed, the dream of a large telescope comes with one caveat. It can consume you.

"You are. You're angling for another telescope."

Facts and figures can be spun a hundred ways and often are. My wife and I discussed the refractor months earlier, when I started building the observatory, and we reached an uneasy truce. But the instrument that I now had in mind was a bit more expensive, and she picked up on that. Telescope creep.

It begins slowly, not unlike buying a car. If fabric is nice, then leather is nicer. Add in the surround sound, the cold-weather package, the all-aluminum wheels and struts, and car salesmen from all parts come running. A telescope purchase is no different. Enhanced coatings, improved tracking, GPS receivers. Soon the children's college fund is thrown to the wind while you try to justify your lack of restraint.

I didn't buy the mobile video link. Don't need it.

A magnanimous gesture that rightfully goes unnoticed. So you're left with this line, groveled more than spoken: "It's the last one, honey. I promise. Really I do."

Some promises are meant to be broken, but the promise of night never is.

The summer constellations are guarding the sky. A year has passed since their appearance, a year spent working and raising children, typing at computer terminals, and otherwise passing the time. But time is precious. It is summer now, not only in the stars above, but in many of our baby-boomer lives. These are the golden years, the stuff of future rocking chairs, Parcheesi, tearful recollections. Someday autumn will come, and winter surely will follow, and someday the stars will come to claim us.

They should. The universe owns our atoms.

It has been said that we are made up of the original Big Bang. The hydrogen atoms that constitute a great deal of our physical makeup haven't changed much in the eons since they were blasted out from the primeval fireball, ex nihilo — from nothing. Only in modern physics can something tangible come from nothing. In the rest of the practical world, nothing from nothing means nothing, as Billy Preston sings. But science doesn't see it that way. Nothing is part and parcel of something, a negative quality of what was, what will be.

Arguments are everywhere concerning the creation of the universe. Most astronomers are firmly positioned in the Big Bang camp, the theory that the universe emerged from a dreadful explosion about 13.7 billion years ago, give or take a few years. The numbers themselves are really meaningless, unless you are Father Time. Time keeps score, though not always. Before anything was, cosmologists inform us, there was also no time, which is an unimaginable concept. Time rushed in with that first ignition, and the clock has been running steadily ever since, as it will do until there is no universe left to measure.

Is that possible? Can the world simply slow, like an old clock left to run down? It is a gloomy thought that even Byron contemplated in 1816, in his poem "Darkness":

I had a dream which was not all a dream
The bright sun was extinguished and the
* stars*
Did wander darkling in the eternal space
Rayless, and pathless, and the icy event

Swung blind and blackening in the moonless air
Morn came and went — and came and brought no day
And men forgot their passions in the dread
Of this new desolation.

Can it really end like that?

No one knows. If there is enough matter in the universe, and the evidence changes from day to day, then the mathematical models suggest that everything could well fall back, reverse itself. The eternal return, right out of Nietzsche, and with every step replayed within the game of infinity. Infinity rules these calculations. It allows for the uncertainty of a universe collapsing upon itself, presumably to Big Bang all over again.

Or the universe could just as easily fade away, with galaxies spaced like desert towns, empty miles between them. Cosmologists worry about this, as well they should. Suppositions about the beginning or end of the universe have a peculiar way of falling back upon themselves, like an Escher print of the paradox of the hand drawing itself portends the universe doing the same, creating and destroying itself at the same time. Even the romance with infinity, in a universe like this, is necessarily tortuous — a marriage that never ends. We are all children of that marriage, hanging around with no other choice but to wonder: Why here? Why now?

My mother, who enjoyed contemplating the astrological origins of the universe, had her own cosmology. She saw the world as a large backdrop to some California movie set. It was all coordinated. Mountains stood in contrast to buildings and stones, each to measure the majesty of the other, while providing the world with color. Large and small, sentient and inert.

"You know," she said, "this exists for us. The universe."

An arrogant view, I always thought, and a tricky one. The tree-in-the-forest argument that loves to tinker with objective reality. Does a tree make a sound when it falls, if no one is there to hear it fall? For the scientist, the unseen is measurable — vibrations of the earth, the infrared blur of a shearing limb, sound waves — but for my mother, these were indications of a reality positioned for human hands.

"All this for people?" I asked.

"Not just people. Every atom, stone, every star. It's here for that."

"Your astrology, I suppose."

"Forget astrology. The universe exists to fulfill a purpose."

"Which is?"

"To simply find a home."

But finding a home isn't always easy. Home can begin in the safety of childhood and end like the mighty Colorado River in Mexico — petering out hundreds of yards from its final destination. The sea.

A confession.

My astronomy career didn't end at the water's edge in Pennsylvania. After high school I moved to Minnesota, during one of the coldest winters on record, to go to college. Minnesota is a tough place for astronomy. A stargazer has to be the hardiest of souls, braving subzero weather to set up his telescope, a trick that I tried a few times before realizing that an observatory wasn't an extravagance in weather like this. It was a necessity.

Minnesotans like their myths. And the mythology of winter creates the greatest stories for amateur astronomers. Eyeballs freeze to glass. Mirrors crack in their cells. Some of these tales, created to maintain a healthy respect for winter, are just urban legends of stargazers who lost corneas like so many tongues pulled from frozen flagpoles. But the point is well taken. It's cold, and the intelligent astronomer needs to consider the harshness of nature. Nature always wins in Minnesota, a discovery I made my first week outside.

It was late December, and I was a new arrival, having bused in on a Greyhound only hours before. Evan, my other brother-in-law, who had been kind enough to let me stay at his house for a few months that became three years, was picking me up at the St. Paul bus terminal. I saw him walking, or rather sloughing, across the floor in heavy boots. He looked like a man fresh off a dogsled: a huge down jacket, with fur around his face; an old stocking cap better left to the Pliocene era; and gloves. Two pairs, the outline of one over the other. He still shivered.

This was in stark contrast to the village idiot. Me. I wore jeans with holes, sneakers, no socks or gloves, and a fringed leather jacket from a rock concert. But I had a telescope in hand, in a cedar carry-on that had my Sears refractor neatly folded and protected from the weather, unlike my common sense.

"Four minutes outside like that," said Evan, looking at me, "and flesh freezes."

The temperature outside was minus

twelve. We ran to the car and drove straight to Target, where he proceeded to generously outfit me in clothes apropos of an Antarctic work station — parkas, moon boots, Thermolite hats and gloves — and we drove to his home. I was greeted with my first northern lights display that evening, a cascading dribble of Day-Glo green, the color of op-art prints and cheap lava lamps.

Thanks to Minnesota's latitude and proximity to magnetic north, the Aurora Borealis, or northern lights, are fairly common. But they weren't common for me. The appearance of such a display right after my arrival wasn't just propitious; it was magical. Solar flares eject particles at Earth with great velocity and hit the ionosphere like iron filings flying to a magnet. What comes next is never the same. Pulsing arcs, clouds, sheets of light, and rays, these auroras appear to fall from the sky in all colors of the spectrum.

Auroras can last for a few minutes, or they can go on and on. I once watched a show for over an hour before going to bed. They also contain a remarkable regenerative power, like embers animated by the wind, flaring up, rolling over, leaping in new arcs, and trundling out different colors. And what colors. I have seen pinks and crimson, vibrant greens, faint yellows, dull and muted purples. They swirl, a palette awash in paint that no earthly artist could possibly mix except in the mind's eye.

"You'll learn to curse them soon," remarked an amateur astronomer from Coon Rapids. "Forget deep-sky photography. Unless you like the added color."

Curse an aurora? I found that hard to imagine.

Minnesota, I discovered, had a strong and tenacious interest in astronomy. There was a large, energetic club in St. Paul associated with the old science center downtown, and several observatories. For the first few months I was enthusiastic, eagerly taking classes while trying to observe in what amounted to a freezer.

It was impossible.

Every night presented unforeseen problems. The first was one of latitude. Minneapolis is higher, by over five degrees, than my childhood site in central Pennsylvania, and the north-polar constellations seemed to loom above me, oppressing my beloved planets. Then there was the weather, which never seemed to improve. Cold, of course. Brutal, savage temperatures that kept an observer on his toes, or made him doubt that he owned any. I rarely spent more than twenty minutes at the telescope without scrambling to the warmth of my sister's kitchen, a kitchen designed to tempt a teenage nose with its pizza, doughnuts, and sizzling burgers.

Why be outside?

And the seeing conditions weren't good. The atmosphere in Minnesota is often unsteady, buoyed by a fickle jet stream. In winter, the jet stream lurches from high-pressure zones — clean Canadian air that kills car batteries, obliterates old starters, freezes pipes. The summer was no better. The jet stream rippled with rising midwestern heat, a rocking and foaming better left to a whitewater kayak course.

Then there were the mosquitoes — the state bird. Ten thousand lakes have bred ten gazillion of these vampires, more than there are stars in the sky, I'm sure. They came in clouds, undulating across bodies of water, from forest and field: hungry, nasty, bloodthirsty. Backyard astronomers in Minnesota have come up with novel solutions to observe. They drape nets around their instruments, with the telescope peeking through. They burn citronella, use bug zappers and toxins. They shower in vats of OFF, eat garlic, build fires just for the smoke. None of this works.

And yet, people still lug their telescopes outside. The desire to stargaze is a strong one, stronger perhaps in dreadful climes than anywhere else. It has to be. Observing is a test of passion and commitment, a test that in less than six months of living in Minnesota I had failed miserably. I gave up on astronomy, thinking that it was too hard. But the hardest part was yet to come.

A quarter of a century later came a phone call. It was during that July of my return to astronomy, and the person on the other end was my father. His voice was terse and high-pitched — unusual for his southern drawl.

"It's your mother," he said. "I got a message. From Seattle."

On television you see an actor's emotions when this scene arrives. There is the obligatory close-up, followed by a reaction shot. Eyes drop, the slim corner of the mouth begins to twitch and almost spasm as the actor's voice plummets into a dark and restless pool. He's speechless. Words fail us in these moments. They lose their power once the shock comes and seldom recover.

". . . human services."

His voice falls an octave. Memory does that. My father lived with my mother for more than thirty years before they split, when I was in college, and he knew her every fiber. Then again, maybe he knew what every husband knows of his wife: only what she wants him to know. The secrets of individuals are deeper than those who love them care to acknowledge.

"The message said to call. It's urgent."

Silence engulfs the receiver. I hear the hum of miles between us. He lives in Pennsylvania but sounds a million light-years away.

"What do you think it means, Dad?"

He didn't answer me at once. But we both knew.

"It could mean anything," he said.

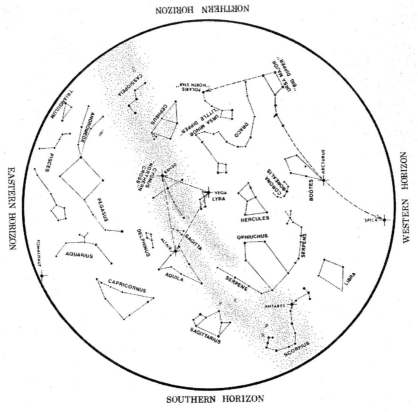

THE NIGHT SKY IN AUGUST

August

Throughout the universe, the geometry of choice seems to be the circle. Stars, moons, and planets are spherical, as though gravity has taken the path of least resistance to offer spatial equality, and the results are round. Life, however, isn't a circle at all, like in the Disney song; it's occasionally an ellipse, an oblong trajectory that sometimes careens far off course only to return, years later, to the same spot. This happens unexpectedly.

A hiker lost in the woods will stumble far off the beaten track to find his bearings. He'll traipse over rock and river, ravine and mountain, only to find another set of footprints. His own.

A comet, fleeing the pull of the sun, attempts one last turn, one final grasp at freedom, before being captured by the invisible glove. Gravity.

Unexpected turns occur in every life. Mine came weeks later in the form of a delivery. It was a telescope.

The call from Seattle wasn't good news.

My mother had died, alone and in her apartment, peacefully. She slipped out of the world more quietly than she had arrived seventy-seven years earlier. She lay on her couch, surrounded by the objects that she loved — her books and manuscripts. Her life was a simple one. Often reclusive, she had little time for social niceties like phones and forwarding addresses, and communication with her was spotty, eccentric, difficult.

She moved to Seattle for reasons that eluded the family. Rain she didn't care for, mountains and ocean even less. But maybe she was ahead of her times, prescient like a good real-estate investor looking for a deal. Seattle would grow up in her years of living there, become the center for grunge music, coffee shops, and software. And I can imagine her, in her black beret, a bohemian on the lookout for the nearest used-book store, latte, intense conversation.

There must have been stars in those conversations. She wouldn't go long without mentioning the stars, or inquiring about their positions in the sky.

"We should check the planets" was a familiar refrain.

My mother never charged for doing someone's horoscope, though she might well have. There were always interested parties, friends who wanted to know about a particular relationship or job offer, and the requests piled up higher than to a call-in radio disc jockey, something I could never understand. Why didn't people simply live their lives and let the chips fall where they would?

The logic of a teenager, or rather the certainty of one, wasn't lost on my mother, who told me that people were searching, and that the heavens were the most obvious place to start. It's the first mystery that most of us witness as children, the great unanswered question of what is out there, in the dark of the great beyond.

"You know that," she said, "or else you wouldn't own a telescope."

"I look up to understand."

"So do astrologers."

"Yes, but I use scientific instruments. Reality, Mom."

"Influence is just as real. The universe exerts a pressure and we react."

This is the gist of astrology. Influence. If science looks for the possibility of a connection, then astrology, having already assumed one, searches for connective meaning. Like a river flowing into its many tributaries, astrology concludes something that science does not: that each tributary flows a particular way because it chooses to, influenced by both source and sea.

This also happens with presidents, I learned. As the scandal of Watergate broke, my mother was busy working on her charts, as a detective would, dusting for clues. She found them. Neptune was positioned in conjunction with Saturn and Jupiter. Worse, it was in Leo.

"Leo rules secrecy, stealth. It's the house of dark secrets."

She whispered, conspiratorially, like a radio-drama character from the '40s. Leo and Neptune, not Watergate, had Nixon by the throat.

"So the stars made him do it?" I asked with sarcasm.

She shook her head.

"No. Certainly there is free will, but when someone does choose, the stars are aligned to confirm their choice, good or bad. Some call this the law of karma. I simply feel the world is in agreement."

"Even if Leo is just a gathering of stars? They change, you know."

I meant the constellations. Both the earth and the stars are moving, and someday Leo himself will be unrecognizable, forcing our descendants into creating new myths, new versions of heaven. My mother knew this, of course. She read constantly; her mind had a thirst that could be slaked by academia, a paradox that I found troubling given her fascination with astrology. Wasn't astrology about superstition, ignorance?

"Astrology is the oldest science," she said. "Ask your astronomers."

I've heard it before. Astronomers like Tycho Brahe were astrologers first, suggesting that astrology gave birth to the science of astronomy, and this may partly be true. But man's fascination with the stars was evident long before that: from the cave, a quick glance skyward offered worlds independent of our own. We are all children of the stars, composed of the same dust and energy, with this exception: from the chaos of creation came a scientific order.

Perhaps my mother knew this. She recognized chaos. It was part and parcel of her world, the chaos and uncertainty of trying to write, and maybe this was her need for mysticism and astrology — to impose order. Like the magnetic poles of an imaginary planet, chaos and control vied with each other, two great and opposing armies threatening a war that they never engaged in. The stars gave her balance, an order that she refused to abandon right up to her death. In her apartment there were recent astrological charts and books. But more important, there was something else.

A telescope.

I'd never known her to be a stargazer. Her looks through the eyepiece of my own Sears refractor were largely ceremonial, cursory glances, as if her world would crumble if she stared any longer. It didn't. Why she got a telescope, I can't say. She bought it around the time of Halley's comet, in 1986. Perhaps she was thinking about our morning sighting of Comet West a decade earlier, and this image could have comforted her, a way of silently connecting with her only son.

Or maybe she too felt the ancient call to look up, push a small glass towards the stars and relish their fossil light. In a world of too much, my mother sought simplicity, and in the end simplicity found her, on a couch in downtown Seattle, three thousand miles away from her children, awaiting the lure of night. It came. Like the words in her journal.

Death isn't the end. It's just a change in direction.

Then why doesn't it feel that way?

There are stars of trepidation and stars of comfort.

After his father died in 1865, Robert

Todd Lincoln, the eldest son of the slain president, took up a hobby that would endure throughout his entire life.

Astronomy.

In 1902, after a successful career as a lawyer in Chicago and later as James Garfield's secretary of war, Lincoln bought several hundred acres of land near Manchester, Vermont, to build his dream house. He also constructed an observatory.

The observatory, a twelve-foot-wide rotating dome located just behind the main house, which he named Hildene, served to protect a six-inch refracting telescope that was made by John Brashear, the famed optician from Pittsburgh. Until his death in 1926, Lincoln used this telescope as often as he could, nearly every clear night, finding a peace and comfort from the stars that eluded him elsewhere. His life wasn't an easy one. Lincoln was present on that fateful evening of April 15, 1865, when his father died, as he would be there to bury his three younger brothers, and later to commit Mary Todd Lincoln, his mother, to a private sanitarium.

The stars offer few answers for a troubled life. They don't even try. But there is an eerie serenity behind the eyepiece, a quiet that soothes jagged souls. It begins in the observatory, sliding a roof open to the heavens. The telescope is always eager for light to embrace glass or mirror, and the evening falls fast. Every backyard stargazer has felt this excitement, the impending rush of night, and the strongest inspiration for building a home observatory is probably to have a spot dedicated only to astronomy.

Not surprisingly, the first private, amateur observatory in America was built in New England, just outside of Boston. In 1823, William Cranch Bond of Dorchester, then an amateur astronomer, constructed a small, wooden observatory to house his telescope and meteorological equipment, instruments that were later used to seed a nascent Harvard Observatory, where Bond would later find himself employed as the country's first paid, full-time astronomer. But perhaps the most famous amateur observatory was located in the unlikeliest of places — on the island of Nantucket.

Born in 1791, William Mitchell was a schoolteacher by profession and later a bank cashier, but his first love was astronomy, an interest that he handed down to all his children. One was especially receptive: a daughter named Maria. Evenings in the Mitchell household were spent looking at the stars, first from a simple observatory built atop the Pacific Bank, where William Mitchell now worked, and later from the widow's perch of the Mitchell house on Vestal Street. The telescope was a small Dollond refractor, an instrument perfect for scanning the skies, which Maria did with great enthusiasm. On the night of October 1, 1847, there was something new in the eyepiece, a faint blur that hadn't been there a few nights earlier. She ran to her father to let him confirm it. He did.

It was a comet.

Maria Mitchell was the first American to discover a comet. She wouldn't be the last. Only a few years after her discovery, in the spring of 1850, a wealthy lawyer named Robert Van Arsdale built a home observatory in his yard in Newark, New Jersey, one much like many amateur observatories that would follow. It had a rotating dome and a four-inch Fitz refractor that Van Arsdale employed for the same purpose that Maria Mitchell had — to look for comets.

Comet hunting is tough and arduous work. It requires long hours at the eyepiece, and even then the process of discovery can be thwarted by a claim, like the stake of gold miners — first come, first rewarded — that can depend on a minute's gap between two astronomers' sightings. This happened to

Maria Mitchell in 1854, when Van Arsdale discovered a comet as she spent time debating exactly what it was.

She wrote in her journal:

Sept. 22, 1854. On the evening of the 18th, while "sweeping," there came into the field the two nebulae in Ursa Major, which I have known for many a year, but which to my surprise now appeared to be three. . . . The bright part of this object was clearly the old nebula — but what was the appendage? Had the nebula suddenly changed? Was it a comet, or was it merely a very fine night?

This is the question that every astronomer must ask himself when faced with a strange object. Is it something new that he's seeing, or something that had always been there but was neglected, never noticed before — in other words, just the result of a very clear night? The eye can be easily tricked, though intuition seldom can, and a comet hunter needs to rely on his heart and intellect for success. Mitchell wrote,

The 19th was cloudy, and the 20th the same, with the variety of occasional breaks, through which I saw the nebula, but not the comet. On the 21st came a circular, and behold Mr. Van Arsdale had seen it on the 13th, but had not been sure of it until the 15th, on account of the clouds. Let the Dutchman have the reward of his sturdier frame and steadier nerves!

This friendly competition, a jostling of wits and envy, pushed these two amateur comet seekers to the edge of their abilities, like a pair of great basketball players vying for control of the same team. Robert Van Arsdale went on to discover four comets, and Maria Mitchell would later be heralded as the first woman astronomer in America. She taught at Vassar College and contributed to the field of mathematical astronomy. But both of their careers began in the only place they could — under the cover of night and in their own simple observatories.

The work on my own observatory slowed. First there was the fog of death, the draining phone calls and final business of a life that had ended, and next the weather chimed in. Rain.

But then my mother's telescope came in the mail.

A cheap blue refractor built by Meade,

hardly larger than Maria Mitchell's first telescope or my Sears refractor, this instrument seemed to answer a restlessness of my own. It wasn't extravagant by any means, but I treated this telescope in a manner that probably wouldn't have surprised my mother. I used it.

She had no idea that I was back into astronomy. At the time of our last communiqué, a few years earlier, I'd been hunched over a desk, not an eyepiece, and there had been no reason since to suspect any change. My mother and I were leading parallel lives of a sort, writing and editing by day, watching the stars at night, though it took me another year to recognize how much our lives had invisibly merged. It began with the telescope. The irony of this gift didn't escape me. What I had once abandoned — the astronomy of my youth — my mother had taken up in my absence, and now it was back in my hands. I was a backyard astronomer again, uniting past with present, the boy with the man.

For two weeks I used that telescope. The circumpolar constellations had now wheeled around Polaris, and the triangular head of Draco, the dragon, was now high in the northern sky. Draco is a wonderful constellation, not just for its mythological connotations but because it's incredibly long, looping its body halfway around the entire pole and back effortlessly. Draco is flexible, like a feline walking along a beam, and it has a seductive eye to match. The Cat's Eye Nebula. NGC 6543 is its official designation. A bright blue puff of smoke, larger than the two stars flanking it, this is a star that came apart at the seams, leaving only a shell. Planetary nebulas are graveyards, the cataclysmic finales of lives spent in the fast lane, and a few are bright enough for even the smallest telescope.

I moved the eyepiece south toward another planetary nebula, M57, in Lyra. Lyra is the lyre of Orpheus, a collapsing box whose strings are plucked by the brightest star overhead, Vega. This is the harp that Nero played as Rome burned, an instrument finer than any amateur musician, much less a madman, deserves. Its strings are said to produce the most beautiful music in the universe, but nobody can hear it. The music is silent as night takes over the withering nebula of Lyra. It evaporates as every gossamer puff must do eventually, into oblivion.

A joker keeps oblivion at bay tonight. He is low on the southern horizon, a shimmering ball of light visible for the first two weeks that August, though it's hide-and-seek for most of the year. It's a game, and the one who plays it, Mercury, is a master. It was said of Copernicus that he never saw Mercury in his lifetime — a myth, no doubt, but one worth mentioning, for Mercury is the most elusive of the inner planets, hardly reaching more than 25 degrees above the horizon. Its fame as a difficult telescopic object is well documented.

Like the moon and Venus, Mercury displays phases as it travels around the sun on an orbit just inside that of our sister planet. We never see a full Venus or Mercury as we do the moon, since that would mean staring through the sun. Phases occur at angles, and a small telescope can let an astronomer gauge the orientation of an orbit with a simple look through the eyepiece.

A few observers have commented on a variety of surface detail on Mercury — more like the seas on the moon, dark shadings — and they have been proved to be accurate. Seen through satellite imaging, Mercury greatly resembles our moon, with its pockmarked landscape of blasted areas, craters, and smooth flows, like a blackened cinder. The temperature on this planet is hellish, hot enough to boil lead, thanks to its close proximity to the sun, and I can imagine no object closer to the sun than Mercury, though others have debated this.

It was called Vulcan.

The idea of an intra-Mercurian planet had been kicking around since antiquity, a kind of stacking toy for the solar system, one orbit inside the other. But astronomers began to seriously consider the possibility in 1859, when a French mathematician named Urbain-Jean-Joseph Leverrier calculated small perturbations, or irregularities, in the orbit of Mercury, measuring less than forty seconds of arc — a tiny amount, to be sure, over a century of observational data. What could have caused this? Leverrier came up with an answer. Another planet.

Sifting through old observations, Leverrier, who now had the audacity to name his undiscovered world, found several drawings that appeared to confirm his ideas. An intra-Mercurian planet would be visible at sunset, or when crossing the sun, he reasoned, as a small dot against the solar photosphere. Unlike the jagged, magnetized edges of sunspots, this would look like a small globe. The planet could also be seen, he thought, during a total solar eclipse — a tiny star near the sun and never escaping very far. Illusory, fleeting.

The word spread, and observers responded. They pulled out their telescopes and charts, as astronomers are wont to do, and spent hours traveling to remote sites to view eclipses, or else they scanned at twilight. But nothing came of it. There were reports of objects crossing the sun that were later discredited, and slowly Vulcan became synonymous with thwarted hope, dashed expectations. But the idea survived. Even today there are occasional searches. And perhaps this is the real allure of Vulcan, that a charred rock could escape detection for centuries, despite solar probes, infrared detectors, and thousands of computer eyes.

We are the sum of our imaginations, not always our discoveries.

There was no Vulcan sighting for me that evening, after the sun had set, but Mercury provided an interesting view. In a telescope, Mercury appears to be leaden in color, like the flimsy ooze of dried-up ceiling paint, but surface detail has always eluded me. Perhaps this is because of aperture size, though more likely it is because of the atmosphere, which is at its thickest near the horizon. The air fumes. And Mercury doesn't stick around long enough to see improvement.

Mercury isn't hard to find, though a good horizon is necessary. Houses, sodium lights, and skyscrapers will inhibit any celestial view, but they're disastrous when you try to observe Mercury. The more sky, the better. Amateur astronomers, frustrated by cities, neighborhoods, and strip malls, have taken to dark sites all across America — fields, farms, stretches of black map at night. But Mercury doesn't need this attention. It only wants clear horizon. It shines at zero magnitude or brighter, and spring is often the best time to observe Mercury in the northern hemisphere, right after sunset. Autumn favors the morning sky and early risers.

Binoculars help. Even an inexpensive telescope, like my mother's, will reveal Mercury's phases. But finding the elusive inner planet is often enough of a prize, one easily bagged with the naked eye. Stargazing doesn't have to be fancy, a shrine to technology and the thirst for more — more telescopes, more magnification, more ability to gather light. It can be as simple as looking up, a lesson that I had forgotten while building the observatory but had relearned with the cheapest instrument possible. Astronomy isn't just about tools, I discovered, it's also about the heart.

Every enthusiast has a shrine. In baseball it's Cooperstown, Montana for fly-fishing, Alaska for hunters, London for book collectors. For backyard stargazers, the shrine is in Vermont, the largest astronomical gathering in the world, at a place called Stellafane.

Every year, in late summer, thousands of astronomers and amateur telescope makers converge on a desolate, windswept hill in southern Vermont, near Springfield, with a zeal usually reserved for religious pilgrimages. Lourdes for astronomers. They come year after year, waves of folks, in vans and cars with personalized license plates — like QUASAR or STARMAN — on expensive motorcycles, driving Beemers, dusty old Volvos, and light-duty trucks. They lug telescopes by the hundreds and chatter in a language all their own.

Many of these faithful are men, but there are some women too. And this year I saw my share of children. There are babies in carriages emblazoned with fiery comets, elementary-school kids wearing astronaut hats and moon shirts, restless teenagers awaiting the ride home or the ring of the cell phone. They come from all over, from parts known and obscure. Kansas, Colorado, Florida, Singapore. Like the relentless pull of a neutron star, Stellafane draws this horde of people stripped of all pretension and jobs. There are no doctors, machinists, or lawyers here. Just astronomers.

It began small, as many great ideas do, with the passion of one man. His name was Russell Porter, and he was born on December 13, 1871, in a house not far from the mountain that would later become famous as a site for gathering astronomers. His upbringing was traditional Yankee, heavy on self-reliance and clever problem-solving. Porter was a man of many skills. He was an Arctic navigator, artist, inventor, amateur composer, house builder, and tinkerer.

He was also a telescope maker.

Telescope making has a long and lively history in the United States. The first American telescope maker to go into business for himself hailed from the Connecticut River Valley, near another Springfield, this one in Massachusetts. In 1806, Amasa Holcomb saw his first solar eclipse. He was 19, and watching the moon slip in front of the sun that day did for him what it does every year to thousands of people: it hooked him. A surveyor and farmer, Holcomb began to build his own instruments, and by 1833, he was selling the first commercial line of telescopes in America, inspiring countless other entrepreneurs to follow.

Success breeds imitation, and soon there were rival telescope makers. One of these, an optical designer by the name of Henry Fitz, also living in Massachusetts, set up shop, and between 1840 and 1855, Fitz had practically cornered the market. Eighty percent of all telescopes made in the United States were designed and manufactured by Fitz, a monopoly that surely wouldn't go unnoticed by the SEC today. But most of the telescopes that Fitz built were small to moderate-sized refractors. For the really great telescopes, institutions like Harvard University needed to go to Europe for their glass, a fact that rankled a young painter named Alvan Clark.

Clark was an artist and an amateur optician. He fooled around with telescopes in his spare time the way people used to whittle, slowly and with a tenacious if careful energy. But everything changed the night he looked through the Harvard refractor, fresh in from Europe. There were errors in the figuring of the lens, he wrote:

> Yet these errors were very small, just enough to leave me in full possession of all the hope and courage needed to give me a start, especially when informed that this object-glass alone cost $12,000.

By 1870, Clark and Sons were supplying

American colleges and universities with large glass, including the famed telescope for the U.S. Naval Observatory, and Alvan Clark's optics were being hailed as among the best in the world. They also inspired a generation of astronomers who couldn't afford his lenses — amateur telescope makers like Russell Porter.

While spending time in the Arctic as a celestial navigator, the young Russell Porter became interested in astronomy, but it wasn't until a friend of his, local business tycoon James Hartness, gave him a copy of *Popular Astronomy* magazine that he found his true calling: building telescopes. The magazine was fortuitous. Inside was an article that changed not only Porter's life but the lives of many stargazers by illustrating what amounted to a lost art, the art of grinding a mirror.

Until then, most amateur astronomers purchased telescopes from commercial makers, and these were expensive, beyond the reach of many. Not so with mirrors. The technology to grind one was cheap, mostly labor, and Russell Porter soon made several mirrors himself, working out the plans for others to follow. As he perfected his techniques he did something that was ahead of its time. He started a user group. The

Springfield Telescope Makers incorporated on August 17, 1920, an organization with sixteen pupils, future disciples in the art of telescope making.

Porter spread the word. He wrote. One of his articles, in 1921, piqued the interest of Albert Ingalls, an editor at *Scientific American*, who visited Breezy Hill, land donated and christened by Porter. There was a house there, and soon an observatory, all built from scratch. Ingalls, an amateur astronomer himself, was fascinated, and he penned a groundbreaking article in 1925, which was followed by a monthly series on telescope making that helped put Springfield on the map. A movement had begun. Students and amateur telescope makers, or ATMs, as they called themselves, soon began to gravitate to Breezy Hill, taking classes, learning about optics and the art of grinding mirrors.

It hasn't stopped.

Stellafane has grown through the years, but it still retains much of its original charm. There are still amateur telescope-making workshops and contests for the best instruments built by hand. Telescope making has evolved much since Russell Porter's day. There are people who construct giant Newtonian reflectors, small refractors, binocular telescopes, fancy computerized mounts. But most of the folks that attempt a telescope on their own do it the old-fashioned way. They grind mirrors.

I'm standing in front of an empty, overturned drum. Around me are other people, perhaps thirty of us in all, hunched over similar drums and pushing a circular piece of glass across a span of plate, roughly the same size. Between the two hunks of glass are abrasives, thin layers of grit and water that cut into the glass, creating a parabolic curve. The motion is smoother than you would expect and the sound softer, more muted, reminiscent of my youngest daughter grinding her teeth at night.

"How long do I have to do this?" I ask.

The man running the workshop resembles the Gorton's fisherman. Big beard, graying and thick, and the upper body of a guy who looks to have wrestled storms, or else pushed a fair amount of heavy glass.

He looks at his watch.

"Dunno. How long have you got?"

Until nightfall, but I didn't tell him that.

"We've been grinding that mirror for months. Let's see how it's going."

He leans down, grabs a bucket, and splashes some water over the glass. Water is important. It keeps the abrasive wet and the two surfaces conducive to friction. Friction helps, digging the abrasive from the tool into the glass, which is now sliding in my hands. He inspects it.

"I'd say another year at this rate."

A year I didn't have.

"The process is a slow one," he says, "but a good glass pusher can figure a mirror within a millionth of an inch. Most accurate instrument on earth made by hand. Imagine that."

But I just kept thinking about that year.

"Aren't there faster methods? A machine —"

His voice is curt. "We don't use machines at Stellafane."

The art of the glass pusher, as ATMs call themselves, is a methodical one. A large mirror, like the one that I had secured on my drum, can easily take twelve months or more to figure, and almost as long to polish. In a world of e-mail, instant messaging, and 24/7 business, mirror grinding is worse than old-fashioned; it's anachronistic. Like calligraphy and gold leafing, the art is in the slowness, the glacial attention to detail.

I look at the other students. A few just shrugged and walked away, not getting it, but surprisingly most of us were still at it.

"You're trying to hurry," he cautions me. "Slow down."

Did the earth hurry to cool? Were the stars anxious to form planets? The process of creation, like the art of a mirror grinder's hands, are deliberate, laborious, well considered.

He illustrates. The motion of the mirror across the tool is back and forth, to and fro, a rocking not unlike putting a child to sleep, setting its own rhythm. Unlike mine. I want to cut glass, make progress, get somewhere.

"This isn't a race, you know."

I push and the mirror comes back naturally. It's dense, a great slab of Pyrex about three inches thick and weighing nearly eighty pounds, but it feels weightless. To the left of me are abrasives in plastic buckets, great heaps of emery, dark and silvery clumps of Carborundum, garnet, cerium oxide — and then there are the polishing agents.

"Wait until we really have to work."

This was the easy part, he said. After a mirror is figured, or ground into the concave curve of a parabola, it has to be polished. This requires a series of steps: making a channeled lap from balsam or sticky tree pitch, heating it to the bottom of the tool. Next comes more rubbing, polishing.

"That's where the hours really go."

Does it ever end?

"Sure. I've finished several mirrors myself. Of course, this is only half the battle. You'll need to test the mirror for imperfections."

"And if it doesn't pass?"

"More grinding and polishing. But eventually it all comes together."

I could feel my own life pass away. Years of grinding and polishing.

"Of course, then you have a mirror but not a telescope."

He signals to the rear of the tent. There was a small woodworking station, and a contingent of men hacking and sawing through plywood, cutting tubes, mounting focusers — constructing the telescope body. One man had been sanding the same spot for the last thirty minutes.

"You'll need a secondary mirror to bounce the light from. Plus an eyepiece and a finder, of course, though I recommend that you buy those."

"Anything else?" I ask wearily.

"Yes." He smiles. "You need to use it."

He was right. Finishing a mirror was only a step in one direction, a mile in a journey of ten. An amateur telescope maker has to be a carpenter, a mathematician, an optical designer, a painter, a machinist, a blacksmith. But most of all, he has to be creative.

Few areas of amateur telescope making have lent themselves to creative thinking more than the mount. Russell Porter was an inspiration here as well, inventing a series of new, revolutionary mounting designs for telescopes, including the cutaway drawings of the horseshoe equatorial at Mount Palomar, which he helped produce. Traditionally, a telescope maker bolted his instrument to a fitted pipe, angled towards the North Star in a great and lopsided **T** that would simulate the motion of the earth. This is called an equatorial mount, and it is popular to this day, though they are manufactured now.

But there's a problem. Mounts bow under the weight of large reflectors, and the greater the aperture size, the more the mount flexes. A 14-inch Newtonian telescope can easily stretch to seven feet long, requiring the sturdiest industrial mount to support it. This once limited amateur aspirations to building only small telescopes.

No longer.

All across the world, at star parties from Peoria to Perth, a new mount has emerged in the last thirty years, one that has solved all the problems of size and weight. It's the brainchild of a passionate stargazer. The Dobson mount.

If Russell Porter was the inspiration for the building of amateur telescopes, then John Dobson is certainly its driving energy. Few people have been so enthusiastic about the telescope, as both art and artifact, as Dobson. Born in China around the time of the First World War, Dobson entered the University of California during the Depression to study mathematics and chemistry, and he taught for a spell before gravitating to a Vedanta monastery in San Francisco. Once there, he turned to another interest of his, one from college — astronomy.

Vows of poverty have no doubt extinguished many burgeoning hobbies, but they did nothing to quell Dobson's passion. He decided to build his own telescope from the simplest and cheapest materials that he could either find or scrounge: cardboard tubes, vinyl records, bottles, sheets of plywood. Soon he was out at night, a telescope under his arm as he made his way to church parking lots, sidewalks, street corners. Anywhere to gather a crowd. His aim was to show people the night sky.

"I want a telescope in every driveway," he once said.

And he did his best to achieve this. He gave his telescopes away, and when they were gone, he started to teach workshops on how people could build their own — cheaply, with materials found around the house. But what revolutionized amateur astronomy wasn't one of Dobson's inexpensive glass mirrors; it was the way that he mounted them. In boxes.

The weight of the mirror stood close to the ground, surrounded by a plywood shell that held the cardboard sonotube firmly in place, anchored yet flexible. A child's pinkie could push these tubes, thanks to strips of Teflon that allowed the box to slide, the telescope to move. Teflon was smooth, slippery. Soon amateurs began to copy his mount, and they began to grind the largest mirrors they could. When I was a boy, a mirror ten inches in diameter was huge for a backyard astronomer. Now mirrors are reaching twenty, thirty inches in diameter or larger, behemoths that require a mobile field observatory just to let them be set up.

And perhaps this is the real legacy of men like Russell Porter and John Dobson. They gave amateurs a way to construct a scope commensurate with their ambitions and interests, and with it a vessel to house their dreams — an observatory of their very own.

September

Summer is quickly leaving. I can see it in the sky at night as the constellations of autumn begin to break above the horizon, a sure enough signal that all is changing. The promise of warm days and lazy nights has given way to the hard work of September. Vacations have ended, and school is in session. People are buckling down again, a schedule imposed on an entire generation of adults, thanks to schooling, that is impossible to shake, even twenty years later. We all feel it. The coming of autumn is serious business.

Perhaps it's ancestral. The first hint of cool air is a harbinger that the circle of seasons will close and everything will return to what it was. Nature seems to know this. Birds are more active, and squirrels scurry around the yard in their final weeks of play, contemplating future hiding places, tiny root cellars for their acorns that now stress leaves and weaken branches. In the forest, there are sounds, the last bleating of locusts that choke the evening. They are so loud that my children close windows just to sleep. But quiet will come. Nature itself is turning toward the final stretch of the calendar year, like an Arabian making a curve in his final race, heart thumping, aging muscles working for one more stab at glory — a glory that comes at night.

Lyra holds the zenith now. Vega, the brightest star in the sky, commandeers the summer triangle overhead and asks what any good leader does of his men, that they obey his orders. The orders of Vega are clear, seasonal. As the point of the triangle, Vega will slowly drift to the west during the course of the evening and the month, dimming summer hope. Follow me, he seems to say.

The great Northern Cross, Cygnus, does. The winged bird leans, nose first, towards the horizon, right behind Lyra. His shape is distinctive, home to a brilliant Milky Way that stretches across the sky, clotted, like clumps of surgical gauze. It's a lost treasure, this galaxy of ours, a white swath that few of us can locate within cities and bright lights. But to our children, the Milky Way has become a sure indication of wilderness — Scout camps, dude ranches, and canoe trips are now defined as much by what can be seen at night, our own galaxy, as by what cannot. Light pollution.

As we return from a long vacation, the signs of neighboring streets and roads elicit joy. We are now home. But at night, especially in summer, home is always overhead. Our galaxy, a pinwheel of stars that is the mother of all we see, gathers us in her bosom. We are never lost within her presence, and the constellations that she holds remind us that the majesty of night is only useful if we look up.

As Cygnus dips to the west, so goes a plethora of objects. A backyard stargazer can spend weeks on Cygnus, looking through binoculars or a telescope, and never get bored. There are novas, stars like SS Cygni that can suddenly brighten by four magnitudes, thanks to a larger, cooler cousin whose hydrogen shell is torn away by a smaller and greedier companion. This has been observed for more than one hundred years, a process of grand larceny that happens on a daily basis in the universe, Darwinian survival except in reverse, the small supplanting the truly gigantic.

Halfway between Vega and Albireo, the bright double star at the nose of the cross, sits a delightful globular cluster, number 56 in Messier's list. M56 can easily be seen through binoculars as another dim ball, though a telescope offers more personality. No two globular clusters are the same. Some are tightly wrapped; others are loose, freewheeling objects that defy resolution, unlike double stars, whose very presence demands a clear outcome in the eyepiece.

Double — or binary — stars are locked in an orbital system; or else they can be optical doubles: masters of illusion. Cygnus is a constellation where these objects occur in spades, from Albireo in the south, on up to the next-brightest star in the spine, which is 17 Cygnus. Like two flames, one golden, the other green, this binary is easily split by a small telescope, but it fools the eye. Two stars where before there was only one.

Multiples rule Cygnus and its environs. There are multiple stars and nebulae. But no object is any finer than M27, the famous Dumbbell Nebula. Most deep-sky targets have handles, like the nicknames that CB radio users give themselves, which offer few clues to their description. M27 is one. In binoculars or a good finder, the nebula appears only as a faint cloud, a mere stain on the glass, but in a telescope the object dramatically fills a low-powered eyepiece. My five-inch telescope pulls out a dim greenish glow with two identical ends, exactly like the weights of a barbell. In a larger instrument, the Dumbbell appears to be twisted in the center, clearly three-dimensional, and quite bright, with extending hoops. The two ends vacillate between looking leaden and looking weightless, not dissimilar to a darkening thunderhead at the end of a summer day.

Finding M27 isn't difficult. It's about a fist length, southeast, from the nose of the bird. But finding the constellation that M27 resides in, Vulpecula, is the real challenge. Vulpecula is an amalgam of the fox and goose, two constellations or groupings really, joined together by nothing less than a wish.

"I can't see it," says my youngest daughter.

The other child huffs. "You're blind."

My voice. "It's supposed to be a fox; I can't tell."

"There's nothing but sky."

"You have to look hard," says the oldest.

"I *am* looking."

The youngest turns to me. She can identify a few constellations easily — the Big Dipper, Orion, and the **W** of Cassiopeia — but she can't find Vulpecula. Nobody can. It's one of the faintest constellations in all the sky, with a few stars that barely stand out.

"Is this a joke constellation?" asks the oldest.

The youngest is indignant. "Why would they tease us like that?"

She takes the star chart and compares it, holding it to the sky.

"It's just a line," she says. "Where's the picture of the fox?"

The oldest shrugs. "Next to the goose."

"You can see it?"

"Can't you?"

That night Vulpecula became the emperor's new clothes. For the briefest moment, I sensed what my mother must have felt when explaining her Rorschach constellations to me as a boy — the power of suggestion.

The power of suggestion, as strong as it is in astrology and in everyday life, rarely finds a home in construction. A builder, not unlike a good scientist, relies on his instruments — straight edge, level, and **L**-square — to measure reality and I was doing the same. Now home from Vermont and inspired, I approached the observatory with a renewed vigor. Already I had two walls up, a pair of skeletal frames braced by two-by-fours. Walls are the easiest of all building projects to construct, and yet there is an elusive quality to them. For walls to be straight, they must also be square. The best way to do this is to measure on a diagonal, from one end to the other, which meant that I needed a lazy finger from my oldest daughter. She looked distracted, bored, or maybe that was just her technique.

"You have to pull," I said, "tighter."

"It is tight."

The metal tape fell over the wall's span like a sagging waistline. And every measurement we made was different. Four pulls, four different numbers. Wasn't mathematics supposed to be exact, absolute? Finally I gave up.

"Let's just nail it to the frame. We'll square later."

She nodded. One eye was fixed on the television, watching *SpongeBob SquarePants*, a cartoon about a hapless creature who was square and yellow. At least something was square.

We were working in the basement now because it was the only room left. The yard was heaped with lumber and cars that had been pulled out of the garage weeks ago. At the houses of most men doing home projects, the garage is the first casualty. In it I had stacked the provisions of a small contracting crew: power saws, hammers, sawhorses, piles of cut lumber and piles of mistakes. There was also dust. Lots of it. This was only half the mess. Hoagies, half-eaten bananas, and spilled soda cans littered the floor, making the place look like a frat party at Home Depot.

It was also nearly impassable. Electrical cords raced from a lone and overworked outlet, spraying a tangled clump out in a dangerous profusion. But miracles abound, as evidenced by the fact that I hadn't blown a circuit or caused a fire yet. There were three saws plugged into one receptacle and an entire salesman's line of battery-charged tools — drills, sanders, and screwdrivers that were always drained and thirsty — into a second.

Kicking a path through, we came in, pushing walls across the driveway on a cart that I designed to save strain on my body, a body whose daily power lift consisted of the morning hoist of a bagel. Unfortunately, all the cart did was provide fun for my children. Our driveway is downhill, a track more perilous than any that Mario Andretti ever imagined, and probably just as fast. Whoops and hollers accompanied my two banshees as they shot past me and braked, Fred Flintstone–style, mere inches away from the walls that I had just nailed together.

"Great cart, Dad!"

Occasionally one would get out and help me with the plywood, gigantic sheets that wobbled in the wind. But soon the sheets were wobbling for another reason.

"Look, I'm surfing!"

My youngest daughter stood on one end and did a Waikiki dance — the fake surfing of old movies. It was the weather. Three months earlier, I had ordered the plywood, an expensive, channeled grade that I was going to cut down to size but didn't. Now the lumber had warped hopelessly, turning up at the edges like bowls, despite being under a tarp all spring. Wasn't lumber waterproof?

No, it isn't. Untreated wood, I quickly discovered, sucks in moisture, acting like a dehumidifier, pulling water from the sky and right into those expensive fibers. The sheets were now heavy. Worse than heavy, they were also dense, resisting the saw blade, which went in about an inch and then choked to a stop. Over and over I tried to cut these sheets, until a cloud of oil and black smoke hung over me.

"You've overloaded the motor."
The man in the orange vest again.
"What did you do?"
"Nothing."

He frowned, an unhappy teacher. I played my part, sounding like a child who was trying to blame the dog for his missing homework.

"It was the plywood. Really, it was."

"What kind of blade are you using?"

Narrow teeth. He sold it to me.

"Let's go with carbide, then. You can hack through steel, if necessary."

I didn't need to cut through steel, just wood.

"Do you have sawhorses?"

"Two." I flashed a receipt.

He nodded deliberately.

"It could be the warping. You know, you can cut this by hand."

Two hundred years ago, he said, there were no power tools. Men went to job sites carting lumber on horse-drawn carriages and caissons, hoisting up supporting beams and walls the only way they could in those days, with brute force. Construction wasn't based upon the seasons, like today, but rather upon the passage of years. A man would start his dream house at twenty hoping to finish it by his thirtieth birthday; or, if he was Thomas Jefferson, he worked on it his entire life. Craftsmanship played a huge part. Using hand tools gave the carpenter time to think, space in which to consider his options as he weighed the cuts of real wood against the patterns left in his imagination.

"Or you could use this."

He held up a jigsaw.

"This baby will do the job. You have clamps, right?"

My expression must have answered his question. I had everything but.

"Well, you'll need some clamps."

Wood, spring, plastic. He dropped these into my cart.

"Lock down the wood. Then try this."

A new blade for the jigsaw. It extended out about four inches, a nasty scar just waiting to happen, and it fell into my cart along with a few spares, a file to shave off the splinters, and more screws.

"You're all set."

I've heard that before. "What should I do with this?"

The dead circular saw.

He shrugged. "Get another. I suggest professional grade."

Twice as much.

"Can I swap it at this register?"

But the man just shook his head.

"Returns. Aisle One."

My mother, when casting an astrological chart, would shut herself up in the bedroom. She needed the quiet. There was the mathematics involved, juxtaposing a person's exact time of birth, to the nearest minute, with the geographical coordinates in latitude and longitude of where the person was born. It all seemed very scientific. A chart was filled with astronomical symbols, jargon that any backyard stargazer could recognize, complete with celestial coordinates like degrees and minutes. Until you looked up. Then it was just stars and the firmament.

The stars on paper are much different from the stars in the sky. In the sky, the constellations tell stories, and nowhere are the tales better than in the late summer. Wars still rage. The battles of Hercules, weakened by his twelve labors, have left him tired, his mind numb. He has slain lions, fought Cerberus. Like all men who have lost loved ones, Hercules can see little point in going on. Perhaps he fights out of rage, or a quest for forgiveness, or maybe because he can rid the world of evil. But evil is vast, and Hercules is just one man.

Just south of Cygnus he finds a weapon — Sagitta, the arrow, a thin row of bright stars that looks exactly like something from an old western, with its long shaft, home to M71, an unusual globular cluster that in my telescope resembles an open cluster, hazy and profuse. With its fletching at one end, Sagitta is lost from the quill of the archer, Sagittarius. Or maybe it's the broken arrow, the sign of war. To the south of Cygnus, Ophiuchus and Scorpio are gang leaders, sending out now Serpens, now Draco, against Hercules.

Above him, to the north, Draco, the dragon, crawls from the cave of Ursa Minor. The bear cub was merely a feint, a diversion to keep Hercules off guard. Now Draco raises his head towards the leg of the hero, whose eyes are on Ophiuchus. Hercules has lifted his club, determined to smash the snake charmer's spell once and for all. But it isn't just Ophiuchus. Half the sky has allied itself in this unholy war: Scorpio, Serpens, Draco, *and* Ophiuchus. One man against the heavens. Can Hercules, the greatest of all warriors, survive?

Even nature flees. Birds whistle from the bulrushes, flying away. The lizard, Lacerta, a creep of stars shaped like two hooked ladles just east of Deneb, slinks for cover. Animals can feel approaching wildfires and wars, and each does its best to avoid such dangers, especially with these odds. Hercules appears surrounded, enemies on all sides, and his doom would be certain if not for one thing. A second arrow flies up from the south.

From the southern horizon comes Sagittarius, the cavalry. A centaur to the Greeks, half man and half horse, the archer rides to the rescue. His bow is flexed, an obvious curve of three stars, with a fourth, an arrow, tugging at the strings and aimed in one direction — toward Serpens and Ophiuchus.

Now the odds are even.

An admission: growing up, I never saw an archer in the sky. My own way of identifying Sagittarius came from the British and is a particularly modern interpretation, stripped of all heroics. The teapot.

There are a handful of constellations that anyone can identify. Orion and the Big Dipper are easy targets, along with Cygnus, the Northern Cross, and the **W** of Cassiopeia, but the teapot is perhaps the most obscure of the recognizable constellations, despite a grouping that looks exactly like a kettle with its pointed lid, handle, and spout. Some of this has to do with the sky conditions in which many of us see Sagittarius, low and in the murky south.

But in a dark sky or through binoculars, Sagittarius can arrest our imaginations. Known for its delightful deep-sky views, dense clusters, and star fields, the teapot has a vantage point better than most constellations. When we look at Sagittarius, we are in reality gazing into the heart of our own galaxy. Through binoculars, great clouds — heaps of gentle dust and stars — contaminate the view. There are more stars in Sagittarius than could ever possibly be counted. Like a bag of sugar chewed into by mice, the slow grains fall and condense into patches, gleaming peaks of white that used to be seen with the naked eye. No longer. Cities and modern lights have robbed the teapot of its primordial majesty.

From my vantage, Sagittarius is impossible to see. It's simply too low. However, over my house and to the south, I can still make out a few of the southern constellations of summer and early fall through a break in the trees. There's Capricornus, the horned goat, looking like a chipped arrowhead. At the point lies the brightest star in Capricornus, Al Giedi, a nice, naked-eye double, though I can't resolve the pair myself.

Capricornus possesses few of the celestial wonders that Sagittarius owns in excess, but it does have the ecliptic running through it, and something else. The outer planets.

As a boy I worked doggedly to bag the two brightest outer planets. Neptune and

Uranus. I remember spending nights outside, with my star charts spread on a card
table, and searching through the finder, a
tiny 5 × 24–power telescope that made Neptune a challenge.

It shouldn't be. At seventh magnitude,
Neptune is a fairly bright object, not quite
naked-eye visibility but easily seen with a
telescope — provided you know where to
look. This is the key. Through a small telescope, Neptune barely reveals itself as a disk
until higher magnification, though it does
emit that telltale glow worthy of a planet.
Planets simply look different. An experienced eye can pick out Uranus and Neptune
in an eyepiece within minutes. Both planets
seem to cast a steady light, devoid of the
scintillation, or twinkle, that betrays most
stars. The light from the planets is warmer.

Advance half a lifetime. I'm back outside,
looking for Neptune again. It's cool, requiring a sweater at night, and I'm sifting
through a pile of computer-generated
charts. On one side of Capricornus, where I
am, is Neptune; on the other end is Uranus.
Neptune won't leave Capricornus for a long
while. Its orbit is painfully slow at 164 years;
it pokes its way around the sun like a tor-
toise crossing a field. Glaciers seem to move
faster.

Because of this distance, Neptune was the last of the gas giants to be discovered — remarkable given its size, roughly five times the diameter of Earth; though its discovery ushered in modern astrophysics, really — a discovery befitting the future of astronomy and talented number-crunchers everywhere.

Neptune was discovered on paper first.

As scientists studied observational reports of Uranus, gathered since its founding in 1781 by William Herschel, an alarming trend was seen. Uranus was falling behind in its predicted orbit, a race car driver with failing tires. Astronomers worried. Newton's Law of Gravitation stated that gravity was a constant, even so far out from the sun. A planet should continue on its merry way, but Uranus wasn't doing that. It was losing precious distance every year — small amounts, of course, but even seconds can add up. What was going on?

In 1845, a young Cambridge student of astronomy named John Couch Adams proposed a theory: Uranus wasn't some renegade lawbreaker, braking on its own, but rather, something was dragging it down. Another planet.

Astronomers, like most scientists, are cautious and rarely jump on a new theory without reliable data. It's the nature of the sciences that truth wins. Until it does, everyone holds their place, at least until someone sticks their neck out.

The first to do this was Leverrier, the French mathematician who championed the imaginary planet, Vulcan, so maybe he had nothing to lose. But this time his calculations were spot-on. He came up with findings similar to those of Adams: that a trans-Uranian planet was influencing the perturbations of Uranus. Leverrier even began to calculate where such a planet might exist. The search was on.

On September 23, 1846, Neptune was discovered in Berlin by J. G. Galle and his student Heinrich d'Arrest, using a nine-inch refractor. It was less than one degree from where Leverrier had predicted — in Capricornus.

Now Neptune had gone full circle.

Thanks to modern star charts, I found Neptune that night with far more ease than Galle or d'Arrest had enjoyed. It is a moderately bright blue "star" sprinkled around distant suns. Otherwise it's disappointing. What keeps backyard stargazers interested in Neptune is simply the same reason that people fish, hunt for bears, raft on white water: for the challenge.

An easier challenge, and one of the most intriguing planets, is Uranus. Located on the eastern edge of Capricornus, Uranus is the seventh-farthest planet from the sun. Though it is at the limits of naked-eye visibility, Uranus was, it seems, unknown to the ancients. All the other planets were already known to the world since time immemorial as "wandering stars." Not Uranus. There may be several reasons for this.

Due to its odd configuration, Uranus doesn't rotate the way most planets do, like a top. Instead, Uranus flails from end to end, pole to pole, like the precarious break of a billiard ball. One pole may even be darker than the other, causing changes in brightness, with a period of prolonged dimming currently going on. Uranus, then, may not have been a naked-eye object all those many years ago.

Through a telescope, Uranus is a frustrating sight. At low magnification a sea-green disk is obvious, but at higher magnification the planet looks like a blurred pea. Few details are visible to small, backyard telescopes, though some observers have seen faint variations in the cloud structure, and moons as well, with instruments over twelve inches in diameter. But the best that most amateurs can hope to do is take note of the brightness of Uranus. Using good star charts and reliable comparison stars, accurate estimates can be made of one of nature's great mysteries: why the light reflected from Uranus can't seem to remain constant.

It may not sound like much. Science is often measured with its small tasks, mundane in their repetition, meager in their scope. But there is nothing meager about the night sky. It reveals its answers to those who watch closely, offering perhaps the best reason to observe with a telescope. The questions to be answered aren't always those of the universe. Sometimes the questions are your own.

Autumn

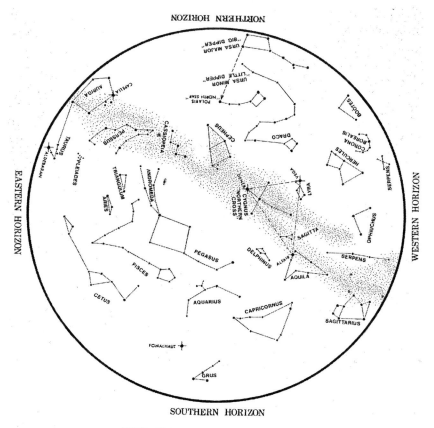

THE NIGHT SKY IN OCTOBER

October

October boasts some of the clearest nights of the year. No wonder. The humidity of summer is long gone and the crisp air has turned leaves orange, red, and yellow as nature begins the process of curling itself up, packing for winter. Here in New England, autumn resembles a tourist postcard. White clouds form the gessoed canvas of a forest heavy with color, color that will thin, empty out, as the month goes on.

The sky too is emptying of summer. Calm patches, like a harbor at night, emerge from just over the eastern horizon, long stretches of blackness that reveal the constellations of fall — Pegasus, Pisces, Aquarius, Andromeda. They converge just beyond the Milky Way, as if stopped by a road divider, unsure whether to cross. On this side of our galaxy, the stars are dimmer, the constellations larger, in stark contrast to the skies of summer, whose battles are now of legend. Hercules sets for another year, a final respite from his labors, and his rest is well earned.

But nature has no respite. She's busy. The great oaks have dumped more acorns than in years past, heaps of crunched and empty shells that squirrels cram into hidden storage bins. Do they know something about winter approaching that I don't? At night the forest rustles. It begins gently, a prattle of leaves encouraged by the wind. I've noticed that the wind picks up in autumn, and the observatory, now a box with four walls, catches everything in the air. Dropping nuts, twigs, leaves. They all collect in places that never existed before this year, in corners, piled on the floor. There's only one way to stop this.

I need a roof.

The man in the orange vest was right. The new saw sliced through my warped plywood easily. Soon I had all four walls up, and they were straight and remarkably level, and I moved on to the roof. The plan was to create sliders. I did this with two 14-foot beams secured to the top of the frame and extending well over the sides. This would support the rafters as they slid off, or so I hoped. In reality, I had no idea. The concepts of weight distribution, dead loads, stress factors, were as foreign to me as tachyons, elusive particles that physicists claim exist even though nobody has seen one. My roof was similar, an idea whose reality was just as frail, even though the gravity exerted on it wasn't.

"Let me get this straight. You want the roof to slide, not crash, right?"

It was the voice of an engineer friend. We were outside, at the last picnic of the year, and I had a sketch of the observatory roof drawn on a napkin for his perusal.

"Then you want cantilevers."

"In English, please."

"A fancy word for support beams."

"I have beams. They stretch along the top."

My friend shook his head.

"You have boards."

"There's a difference?"

"Structurally there is. Wood flexes."

"That's good."

"It's bad," he said.

Wood can fail. A bent match doesn't curve much before snapping.

"What you need are angles. Cantilevers."

Then he described what he was thinking. Four angled supports would attach to the frame, holding both the beams and garage-door hardware in place.

"But not all the way to the bottom," he warned. "That's too far."

For a weight to be evenly distributed, he

said, it needed to have a center of gravity, a balance point between resisting gravity and falling.

"You need to find this."

He sounded like the old monk in *Kung Fu*.

"How?"

"Well, you could calculate it."

I couldn't calculate my own checking account.

"It's actually pretty easy."

He took the napkin and began writing in an esoteric code. Mathematics. This is one of the many reasons why I'm not a scientist. When faced with calculations beyond first-grade math, I seize up. For every formula I see a myriad of possibilities, a thousand answers. It's both the curse and blessing of high school math tests. You can always crank out some sort of response, especially a wrong one.

"Or you could go right here," he said, looking at my glazed eyes.

I took the napkin and folded it.

"Test it first. Buildings have fallen on less."

But not this one.

The sun shines every day now. While I construct the bottom part of my roof, a wooden shell that will support the rafters and wheels, the last rays of summer beat down from an autumn sky.

Like the moon, the sun weaves up and down in the heavens, a drunken sailor on shore leave. Its path varies because the tilt of the earth varies, leaning away from the sun in winter, towards it in summer. But now the sun is beginning to ride lower in the sky. The days are changing, and I can see it when I look up.

As a boy, I took great pleasure in observing the sun. A backyard astronomer can do this as well, despite the dangers of looking at the sun directly. Every telescope comes with a sticker and the attending legalese, designed to keep telescope manufacturers out of court.

Warning! Never point a telescope at the sun.

Good advice.

The sun can blind you in seconds. Large optical observatories, fearing the worst, have computerized controls that prevent an operator from accidentally pointing his instrument anywhere close to our nearest star, and for good reason. The heat collected from a large mirror conjures dangers that only a science-fiction writer can appreciate, such as death rays, instant barbecue. Manufacturers, therefore, have built specialty filters for observing the sun. Each is coated with a kind of reflective Mylar or metallic glass that fits over the aperture end of the telescope, not in the eyepiece. Eyepiece filters have a nasty habit of cracking from focused solar rays and should be tossed in the trash.

A safer way of observing the sun is to project the light on a piece of white cardboard. As a kid, I did exactly that, making money in the process. I recall cleaning up neighborhood coffers during one partial solar eclipse by projecting the eclipsing sun on one of my mother's bedsheets, which I'd strung up like a theater screen as my sister peddled tickets.

Observing the sun is unlike any observation that a stargazer makes at night. For one thing, the sun is our closest star, and the most available for study. Behavior on our own sun gives us an insight into the life cycles of other stars, and many of the activities on our own star, such as sunspots, solar flares, and coronal activity, have been registered elsewhere in the universe.

Our sun is a medium-sized star, nothing special, except for being the supplier of nearly everything that makes life possible on earth — warmth and light. Earth is situated at the perfect point, it seems, at the only place in our solar system for human beings to have evolved, a confirmation of dumb luck or design, depending upon your perspective.

For my mother, the perspective was always one of intent. Her anthropic cosmology, the belief that the world exists for human eyes, was bred from an astrological understanding that the heavens are mirrors of earthly toil. The theory is an ancient one, and interesting; though it counters much of the work that astronomers do every day, observing processes beyond human control.

"Why are we here and not somewhere else?"

The boy in me shrugged. The astronomer in me did not.

"Notice," she said, "that we live on a world perfect for us."

For that, we can thank the sun. A delicate blend of distance and primordial elements have contributed to life on Earth. Hydrogen is abundant in the universe, but it is wonderfully abundant, in a more complicated form, on Earth, as water. An orbit too close to the sun and the water would evaporate, creating more clouds, trapping in deadly greenhouse gases, and choking off anything that walks, crawls, or slithers. Too far away and the water freezes, or gets stripped from the atmosphere like what may have happened on

Mars. For most of us the sun is a friend. But it can also be an adversary.

The visible surface of the sun, called the photosphere, is home to an ever-changing array of sunspots, variations in the magnetic fabric that look like polarized metal filings. The first astronomer to observe sunspots was Galileo, yet there is some evidence to suggest that the ancients knew about these as well, or at least that they saw periods when the sun had a defect. These spots are black in the center, in an area known as the umbra, with a lighter shell called the penumbra around the edge. A good dermatologist might recognize these as moles going bad, cancerous, for they appear that way, but on the sun these spots are a common event, peaking every eleven years at solar maximum.

Often these are flanked by bright areas, called faculae, that float high above the photosphere. Faculae seem to appear before the creation of a sunspot and hang around long after it's gone, wearing out their welcome. Or maybe they're just harbingers of the worst that the sun has to offer.

The solar flare.

Solar flares are massive eruptions near or within a sunspot group. Sunspots like company, and they are usually seen in gangs, hanging out, or recruiting more into the fold. The larger the group, the more perilous the rip in the magnetic fabric, making a solar event more likely. A flare is the energy bursting forth, the pent-up nuclear frustration that bombards our planet with a variety of deadly cosmic rays, knocking out communication satellites, disrupting broadcasts, and irradiating frail bodies.

On March 13, 1989, a blackout in Quebec was blamed on a large solar flare. Although this wasn't the big one, scientists who worry about the great earthquake or an asteroid strike are also concerned about the mother of all solar flares coming. They expect an eruption so large that it could have originated from the mind of a fantasy writer rather than Mother Nature. If this were to happen, electronics all over the world would be fused, some scientists say, bombarding Earth with radiation more lethal than that at Hiroshima.

It isn't a perfect world, by a long shot.

All throughout the summer, there have been smaller flares and large sunspot groups as well. One of these sunspots, a stain that has lasted for almost a month, is so massive that I can see it with my naked eye.

"Is that real?" asks my youngest daughter.

She looks at the white card that I'm using

to project the sun. The image is large and slightly mottled, with the sunspot group scattered across the solar face the way dust settles on my keyboard after a long absence.

"That's a bug, Daddy."

I assure her that it isn't.

"Looks like one. A big fly."

"Those are spots on the sun."

She touches the card, smoothes her finger over the sunspot.

"Are you going to start taking your telescope out in the daytime too?"

It's a question that I often ask myself.

Since my return to amateur astronomy I've noticed the cascading amount of equipment now being offered by manufacturers and clever optical engineers. But none of it is more exciting to me than an instrument I can faintly recall from my teen years — the hydrogen-alpha filter.

Viewed in regular white light, the broad spectrum of everyday, humdrum existence, the sun shows an array of interesting features: sunspots, bright faculae, some faint granulation. But to really see the sun properly requires the use of another part of the spectrum, one especially attuned to a narrow line called the hydrogen-alpha band.

Here the sun explodes in full atomic glory. On the limb, great and arcing glows of gas, called prominences, are cast out like fishing lures, cloudy filaments that battle gravity and struggle to break free of the sun. Some do for an hour or so. Changing, growing, these prominences are often many times the size of the earth, before the sun's gravity finally reels them back into the field.

My first look through one of these instruments stunned me. I could see granulation, like scarlet Rice Krispies, spotted across the entire solar surface, and sunspots, normally frozen in their metallic black, that seemed to swirl, come to life in dreadful agony. Astronomy was clearly more than just observing at night. There was the sun, which most amateurs take for granted, though a few solar observers never do, and their dedication contributes to our understanding of our closest star.

Dedication is at the core of the hobbyist. A mountain climber learns about weather, physical conditioning, geology, and a hundred other things in order to conquer a mountain. He also has to study. So it is with the amateur astronomer. It isn't enough to know a smattering about optics, physical science, or even mathematics; an observer must coordinate this information to make it meaningful. Just as important, he must also stay current within his chosen field.

Many astronomers, even casual backyard ones, have a particular interest. They may enjoy studying the sun, or the planets, or perhaps they are drawn to the variations of pulsating stars. It doesn't matter. Passion drives the art. Soon an observer is reading, studying the professional journals, and learning everything he can. Next, he might join a club or group.

Astronomy clubs have multiplied since I was a kid. I recall trying to start one at the local museum with all the verve and frustration of a teenager. Like most children, I wanted instant gratification — lectures, sprawling telescopes on the grounds, star parties — without doing any work. I wrote a few articles for the newspaper, and I pawned the idea off on the museum director, who was probably overwhelmed and underfunded. Then I sat back to watch it happen. Nothing did.

But thirty years later, there is a club in my hometown, and there are astrophotographers as well. They have wings, these backyard stargazers, and have taken flight to every part of the community, spinning off chapters, lecturing to elementary-school-age children, operating public telescopes. I take no credit for this, of course. Astronomy clubs are a natural inclination for people who surround themselves with folks of similar interests. They also illustrate the health and growth of amateur astronomy. Astronomy is alive and well. Moreover, it is thriving.

Some groups, though, have always been with us.

Since 1911, the American Association of Variable Star Observers in Cambridge, Massachusetts, has acted as a sort of astronomical clearinghouse for observational data pertaining to the changes in the brightness of stars. Many stars exhibit a defined pattern, or light curve, and there are fleets of interested amateurs around the world — armchair scientists, really — adding to our stellar knowledge by observing these distant suns and recording what they see on a nightly basis. The information is then used by working astronomers as a kind of heads-up, a method of signaling to the professional community that something is happening and of value to watch.

This was the dream of a writer and lawyer, William Tyler Olcott, who, almost a century ago, appealed to a loose gathering of backyard astronomers with this call to arms:

It is a fact that only by the observation of variable stars can the amateur turn his modest equipment to practical use, and further to any great extent the pursuit of knowledge in its application to the noblest of the sciences.

Today, the AAVSO has mushroomed into the network of observers in more than fifty countries and counting. And with observations growing by a half million or more each year, the vision of William Tyler Olcott seems prescient. Science has advanced.

Other stargazers are interested in events closer to home, like planets. The Association of Lunar and Planetary Observers, a group modeled closely on the venerable British Astronomical Association, was founded in America in 1947 by Walter Haas. ALPO is dedicated to the advancement of long-term, systematic study of the planets and the moon. Drawings and photographs are pooled, and if anything new happens on one of the planets, chances are that an ALPO member will be the first to notice it.

Science benefits from these amateur efforts. There are teams of passionate observers scouring the heavens for comets; determining the orbits of asteroids; looking for sudden nova, or exploding stars. The amateur contributes to science every day in small ways as science contributes to our lives in larger ones. And despite the advancements in planetary probes, space telescopes, and optics, professional astronomers are still a rare breed, with dwindling numbers. There simply aren't enough professional eyes to keep tabs on the universe, so amateurs pick up the slack.

This is nothing new. The history of early American astronomy is really a history of amateur astronomy. Then it changed. In 1845, William Bond, who was already working at the Harvard observatory without pay, was suddenly offered an actual salary. Slowly the professionalism of astronomy grew, and the final split came in 1895, when two magazines, *Popular Astronomy* and *Astrophysical Journal*, divided the market. The message was clear. There was a popular hobby, and then there was real science.

For years after that, amateurs did their best to join the professional ranks. They purchased telescopes, built observatories, made observations. Their work was serious, and sometimes it paid off. Clyde Tombaugh, the patron saint of backyard stargazers, was an amateur astronomer himself when he walked off a small Kansas farm and into Lowell observatory. His job was a minor one, and seemingly perfect for an amateur.

Search for Planet X.

The mathematician and astronomer Percival Lowell had long speculated about a ninth planet, one lying far outside the orbit of Neptune, and it was his dream to have his beloved Planet X discovered in his own observatory. A small instrument was dedicated to the task, one that could take pictures. Photographs were necessary to find such a dim object. Planet X would be faint — around 13th magnitude, Lowell said — and its motion would be tiny. The only way to find it would be to compare two images, taken a week apart, and measure each individual star.

Every night, Clyde Tombaugh spent cold evenings at the telescope, photographing long into the morning. Then, after a few hours of sleep, he would sit all day at a blink comparator, an apparatus with binocular eyepieces, and he would stare at the two plates. The photographs were of the same star field, and by blinking them on and off, right and left, Tombaugh could detect any motion from the stars as light slalomed back and forth.

He did this for months.

One day, he simply found it. And now schoolchildren all over the world know the name of the American astronomer who discovered the planet Pluto.

Clyde Tombaugh represents the best of amateur astronomy. Despite advancements in telescopes and knowledge, what drives an observer most is his dedication, his love of the night, and that love transcends all else. This love also has the power to heal — not just broken hearts, but it can, in the words of the first woman astronomer, Maria Mitchell, "bring calm to the troubled spirit and hope to the despondent. When we are chaffed and fretted by small cares, a look at the stars will save us from the littleness of our own interests."

The littleness of my own interests recede with the call of night. Solar telescopes and big glass — these are only tools, like the observatory, to embrace the stars.

Above me, high overhead, I see the Milky Way. The tiny constellation of the Dolphin, Delphinus, a neat, diamond-shaped asterism of six stars, takes its turn with the observatory, where I now find myself sitting. Gamma Delphini, at the tip, is one of my favorite double stars — two golden sparklers of equal magnitude separated by a whisper of black. Like diamond earrings on an elegant woman, Gamma Delphini is simply riveting, and I never miss a glance.

Tonight I am sifting through the Dolphin for more elusive quarry, the obscure galactic cluster NGC 7006. Cataloged more than one hundred years ago, this faint object is located to the east of Gamma Delphini, a faint blur in my telescope. When I do locate it, not with the finder, but by sweeping slowly with a low-powered eyepiece, I am surprised. It looks brighter than I thought it would, and has a tarnished sheen of silver.

Just east of Delphinus, I can see a great square rising through the trees. By 11 p.m. it is almost overhead, a neat geometric form that now reveals its full presence as Pegasus, the winged horse. Pegasus is an unusual constellation; it's two constellations, really. At the corners of the great square are four bright stars, easily seen in any city; these are four points that any stargazer can identify: the red star Scheat to the northwest; Markab, the brightest, to the south; Algenib, just west; and Alpheratz to the northeast — which is actually the brightest star in the constellation of Andromeda.

Why the confusion?

I look up. As Hercules sets in the west, a new set of constellations and stories emerges. The sky is built upon drama, and autumn is no different. The wars of Hercules have ended. They are now tales, told in the night between the seasons. But the sky is not calm.

Water has replaced the land. There is Pisces, the fishes, just southeast of Pegasus, a school that swims up a dangerous and troubled current. Like the famous depths of Loch Ness, the water is murky, opaque, but a form is seen, or rather the ripple of a gigantic fin. This is Cetus, the Whale. Unlike the friendly, endangered whales of SeaWorld, Cetus is a familiar of Melville's. He is Moby-Dick, the great Leviathan, destroyer of human spirit, nature run amok.

Cetus plumbs the depths with a wary and careful eye. He is a realist, a believer in the material world — unlike his magical adversaries, who rely on stealth and incantations; women who can turn men to stone. Not Cetus. Girth is his reality, the density of water, the pressure from the deep that he uses to his advantage. He strikes at a moment of his choosing.

The moment is now.

On a lonely and barren island, waves batter the shore. There are no trees on this windswept scrub, only rocks, and something else. A woman.

The woman is a beautiful maiden named Andromeda. She is chained to the rocks, a helpless form. Her legs are folded; her eyes are downcast. Andromeda is doomed, and she knows it. Hunger incites the whale, and Cetus circles the island, planning his mode of attack. He could smash the rocks with his tail, or perhaps he could just leap up and beach himself, as every whale knows how to do for occasions exactly like this.

Andromeda weeps. Who wouldn't? Like the sorry soul in the Edvard Munch painting *The Scream*, Andromeda wails for justice but there is none.

Her mouth is the outline of the Andromeda galaxy. I have seen this object with the naked eye many times, often late at night. In a telescope, the Andromeda galaxy, known as M31 to astronomers, swallows up most of the eyepiece. It's huge. The central core, a ball of diffuse light, reveals the twist of spiral arms that in a long-exposure photograph betrays its significance. This galaxy closely resembles our own.

Finding M31 isn't hard, especially with binoculars. They are a useful tool for beginning astronomers, and I would recommend a pair for any novice observer who wants to learn the sky. Sky knowledge is important. It shortens the time between searching and actual observing, and it can offer a lifetime of pleasure. Knowledge also helps when finding faint objects like M31.

Begin at Alpheratz, the northeastern corner of the square of Pegasus. Continue northeast to the second-brightest star in the knee of Andromeda, Mirach, and move northwest about a fist length. You can't miss it.

Finding Mirach also allows an observer to locate two more targets. One of these, Gamma Andromedae, also called Almach, is the brightest star up from Mirach. Almach is a classic double star, delightful in small telescopes, two Christmas-tree lights of yellow and blue. But Almach's companion is also a double, with an orbital period of sixty-one years, and the separation of these two stars requires the optics of a larger telescope.

The other sight near Mirach is to the east, in the constellation of Triangulum, the triangle, which really does resemble an architect's triangle. The object is M33, the Pinwheel galaxy. Face-on, M33 looks exactly like a pinwheel on a stick, its spiral arms looping in nice syncopation. But few other targets have caused so much debate when seen through a telescope. Many observers complain that they can't find this galaxy, although it is large, almost filling a low-powered eyepiece, and relatively bright, while a few claim the opposite — that it's so easy to see that they have actually found it with the naked eye. I am between the two camps, having seen M33 easily in a finder though never while looking straight up.

Here is a trick that many experienced stargazers use when looking for faint objects: try to look at the object sideways, to the left or right, rather than directly. The retina is more sensitive on an angle, and using this averted vision helps to increase the natural optics of the eye. Dark adaptation helps. The eye needs about thirty minutes outside for maximum performance, away from televisions and even flashlights. Astronomers use red lights to preserve their night adaptation, though even a bright-red light will affect the eye. The best trick is to stay in the dark.

The eye adapts, and soon the world opens up. My forest is never as dark after an hour outside as it is when I've just walked out. I can discern details. And the sky, after the same hour of darkness, is never as ink-black as I first imagine. Gray seeps in. Soon the sky doesn't even look all that dark.

An amateur astronomer will do anything he can to preserve his dark adaptation.

Blinds around the house. Lights off. Others opt for less conventional methods. They put up fabric blocks over neighboring street lamps. A few even go as far as possible, as far as I went. They build observatories.

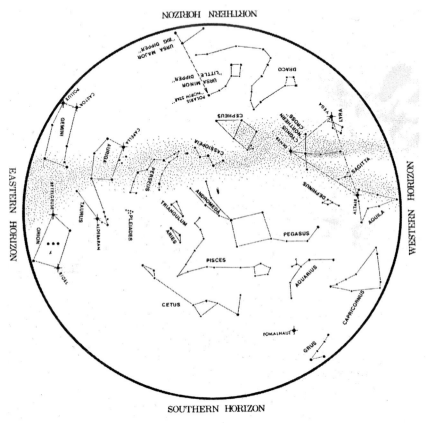

THE NIGHT SKY IN NOVEMBER

298

November

The stars of November foretell a future chill. Winter is coming.

You can smell the night. A trail of wood smoke lingers in the air, thin and nebulous, rising high above the trees, a sign that temperatures are falling. But a night brilliant with stars also has an odor, as distinct as bread baking in an oven, as crisp and scrubbed as newly fallen snow. It's the aroma of freshly cut applewood. Almost sweet, it hangs in the air as Christmas stockings do over the mantel, inspiring great anticipation.

It is clear tonight, and the air is cold. I'm wearing a short coat, collar up to trap my body heat, and gloves. All the trees have been stripped of leaves, and the forest behind the observatory looks barren. A poor layer of frost dashes across the yard, and it would scamper across the telescope optics as well if not for my heater. A wire, about the thickness of dental floss, sends a weak current of electricity through the corrector plate, keeping it warm and frost-free. Refrigerator technology for the battlefield of stargazing.

The stars twinkle as the atmosphere settles in, revealing the constellations of fall. Throughout human history, the sky has been a theater, providing the visual backdrop for stories of triumph and failure, dramas that are retold with each passing season. Above my head roams an assortment of creatures, some imaginary, like the winged horse of Pegasus, others real, like the giraffe, Camelopardalis.

This giraffe has long, spindly legs and neck, and she is defending herself. From the northeast comes a trail of stars, camouflaged in the brush, as tigers often are. This is Lynx, a predator stalking near the Pole Star for game. Every year the herd returns, as predictable as an eclipse, and every year the predators wait. Soon the savannahs are filled. There is Aries, the bighorn sheep, just below the triangle, and the animals around the rest of the sky are great in number, bear and waterfowl, so plentiful that they attract the largest predator of all. Man.

The man is a king, Cepheus, and he brings with him a queen, Cassiopeia, whose distinctive grouping of stars, a **W** just south of Polaris, sticks out like the jagged grin of a

Halloween pumpkin. This is my oldest daughter's favorite constellation. It's the name that attracts her, being phonetically similar to her own, but it's the shape that causes her to study it. She likes the giant **W** in the sky. It never looked like a queen to me, either; Cassiopeia was always the chair. A throne.

A king needs a throne almost as much as he needs a queen. Sitting next to his queen, Cepheus is a distinctive constellation in his own right, but for reasons more pedestrian than Cassiopeia's. He looks exactly like a house. With its peaked roof and squat foundation, this home is found in the drawings of school-aged children. There are five major stars comprising the house, but it is the one to the east of the bottom star that shocks everyone. This is Delta Cephei, the fourth-brightest star in the constellation and one of the most famous objects in the entire sky. In 1784, the amateur astronomer John Goodricke first noticed variations in the brightness of this star, and it was later discovered that Delta Cephei pulsates. This caused the designation of an entirely new class of star: the Cepheid variable.

Cepheids are giant stars that throb with a frightening, clocklike accuracy. They have waxing-and-waning periods as short as a few hours and as long as 50 days, though many Cepheids establish their light curves within a week. An amateur can observe these stars and note that their magnitude changes, often in relation to other, surrounding targets, and then calculate exactly how bright these stars are. Once graphed, the entire picture comes into focus — a repeating curve.

Some of these Cepheid variables have been observed, night in and night out, for more than two hundred years. Goodricke discovered three himself and helped to open up the field of variable-star astronomy, though he wouldn't take it very far. He died, at the age of 22, after falling ill during an observing session — a danger to all astronomers, catching their death of cold outside — but the stars of Goodricke live on.

As do the constellations. In the November sky, both Cepheus and Cassiopeia reach their farthest points toward the zenith, having rotated all year long around the pole and out of the trees. The **W** of Cassiopeia now stretches above my worst offender, an emaciated oak that wants to hold on to its leaves as long as possible and struggles to dominate the sky. It can't. Cassiopeia's chair consists of six very bright stars, obvious even through branches or in the harshest city light. Through a telescope or binoculars, the field is starry, almost foamy white, with the river of the Milky Way tumbling its way down from Cygnus.

An amateur could spend months exploring Cassiopeia. Robert Burnham Jr., in his delightful *Celestial Handbook*, fills almost sixty pages describing the cornucopia of sights that Cassiopeia makes available to the stargazer: variable stars, open clusters, binaries, galaxies, diffuse nebulae, radio sources, and one of the most exciting events of the past five hundred years — Tycho's star.

This is the great supernova of 1572, discovered by the Danish astronomer Tycho Brahe, who looked up at Cassiopeia, just north of the bow in the **W**, and found something that didn't exist before. A new star. Over the next several weeks, the star grew brighter until it became the equal of Venus, an object so brilliant that it can sometimes be seen in the daytime.

The death of stars is as varied as that of people. Old age kills most. Stars swell up, exhausting all their nuclear fuel, their life force in the process, and then retreat back to a state before birth — that of a white dwarf. Some massive stars shed their hydrogen shells only to become superdense dwarfs called neutron stars. Others collapse beyond the neutron-star phase and become black holes.

But a few stars experience cataclysmic failure. Like a weakened aorta, they simply burst. These are stars on the gurney, the emergency-room cases that get tagged, the superstar basketball player dropping suddenly on the court. For reasons unknown, the fusion cores of these stars become unstable and they explode, sending a torrent of light in every direction. In 1572, Tycho witnessed a similar explosion that remained visible to the naked eye for sixteen months.

Now Tycho's star is a memory. You can't see it in a telescope, though the Palomar observatory has photographed faint cloudy remnants. These are not nearly as well defined as is the Crab Nebula in Taurus, a puff of eerie green smoke, as if blasted from a cannon, that hangs in the eyepiece. Nor is Tycho's star the stuff of the Veil Nebula in Cygnus, the faint exhaust from a dying engine. The remains of supernovas are fairly common. There is Barnard's loop, a photographic ring around the edges of Orion, that suggest an ancient cataclysm, and another cloud in Gemini.

Some exploding stars leave a signature.

It's a signal that can be picked up by radio telescopes, instruments that use powerful antennas to listen in on the cosmos and detect patterns. Like an echoing cry through space, these reverberations from the past offer the hope that some night an amateur will look skyward and find a brilliant new star. It's happened before. It could happen again.

But not today. The sky is spitting snow, and from icy nostrils comes exhaust, like the sputter of an automobile, except the automobile is me. I'm wearing gloves and pounding nails, working on the roof. The garage-door hardware is installed, and I'm finished with the sliders, a mechanism that will allow the roof to glide, separate in two halves, except that I have yet to try it.

The man in the orange vest, an hour earlier, knew as much.

"Do you have any idea how much a roof weighs?"

I did, but he showed me anyway, handing me a heavy block of shingles.

"You'll need three bags, plus plywood. Half-inch. Anything less and it buckles. A lot of pressure on those wheels, sport."

He was right. I had no idea how much weight the tracks could withstand. Garage doors were one thing, a heavy roof with snow another.

"Of course, you could go without shingles."

He pulls out a sheet of new material. It was green.

"They love this stuff in the islands. Very light."

"Plastic?"

"Corrugated roofing."

"You mean plastic."

Sure, it looked exactly like plastic, he said, but nothing was easier to work with. Simply lay it out, cut the sheets with heavy shears, and nail the roof on.

"It's also cheaper."

The magic words. I came home with lumber and a new roof strapped to the car, which had begun to resemble a mobile hardware store: cords, saws, and plans. The observatory had not only taken over the yard and portions of the garage; it was also in danger of taking over everything around the house, including the family cars. Wives notice the creep of building projects that are getting out of control.

"Shouldn't you clean up as you go along?" she asked innocently.

But the pile of junk in the garage suggested otherwise. I was closing in, with all the obsession and focus of a guy running a race against himself.

"I'm almost done. Just the roof to go."

Roofs like this one are best built on the ground.

Once I had the rafters up, the roof began to resemble an Amish barn raising — an eerie skeleton on wheels, ready for its plastic flesh.

The material that I had selected, a green composite more often found on roofs in the Caribbean than in New England, was exactly what the man in the orange vest had promised. It went on easily with rubber-shanked nails that kept water out of the breached plastic — and breached it was, by small nail holes that I caulked as I went along. But the roof was a breeze, which should have been a warning, and soon I was hoisting it up, this time with help.

There were three of us. I'd called a good friend of mine up the road for some assistance, a flexing of muscles that neither one of us was used to. Answering the phone is dangerous in America. You can be asked to do anything, like lift an observatory roof. Fortunately my friend brought an extra hand, his brother-in-law.

Over the past few months I had developed a spiel that I gave to the curious, a long-winded speech on what I was doing and why, but it wasn't necessary with my friend's brother-in-law. He knew just from looking.

"My boss built one of these," he said non-chalantly.

Imagine my shock. "An observatory?"

"Yeah. You should see it. He has a warming room and a bed."

"He has a bed?"

"A man has to sleep sometimes."

He glanced inside my little hut.

"There's a TV and a satellite dish too."

I felt myself beginning to shrink. The problem with comparing observatories, not unlike comparing fancy cars and expensive audio equipment, is that under the experienced scrutiny of male eyes, every purchase falls short. So had my observatory.

"He really has a satellite dish?"

"And GPS. It's solar powered, the roof. You'd love the setup."

I nodded like I was going to my execution.

"The whole deal is computer driven. Even the security."

"Security?"

"For the telescope. It's huge. Cost more than my car."

"That's some observatory."

"Oh, it's a beaut. He spends every night

out there. I don't know what his wife thinks."

There's a fine line between passion and obsession. At the heart of love lie the potential seeds of destruction. You notice it in many hobbies. The lepidopterist going on one last trip for that elusive butterfly. The kayaker determined to make a final stab at a Class V rapid. What makes us secure in our own interests can just as easily warp them — when we go too far, or perhaps not far enough.

I study my own humble observatory with new eyes. Why didn't I build a larger shed again?

"The guy even has a video link. Some observatory in California."

"Mount Wilson?"

"Maybe. Anyway, he can link up to other telescopes online."

"Geez."

"It's nuts. But don't get me wrong. This is . . . nice."

It was nice, sure. Fondue nice. Yugo nice.

For the next several evenings it was hard to go outside with any enthusiasm. Just hearing about this fancy observatory was enough to deflate me. What an old garden shed to the world of science?

Howard Carter was an artist and tour guide before he became an archaeologist, traveling to Egypt and discovering the tomb of King Tutankhamen. Richard Leakey was a high school dropout when he led his first dig. These are a few examples of amateurs who made a splash in science. But a backyard stargazer, despite the possibilities of meaningful scientific work, has to measure his own accomplishments in smaller bites, with the passage of time.

An amateur astronomer's greatest asset is time. Over the course of many years, long-term systematic work can be scientifically useful, work that professional astronomers can't often accomplish. They're squeezed by teaching, budgets, writing, and valuable telescope time, so the heavy lifting goes to the backyard stargazer. He measures the light curves of bright variable stars, the weather on Jupiter, the orbits of asteroids, with hardly a complaint. It's fun — at least until he begins comparison science with the pros.

What brought me out of the doldrums that night began as a wisp, a fleeting speck that appeared and vanished so quickly that I thought at first it was an illusion.

It wasn't.

A second one came, three minutes after the first. But this star left a mark, a phosphorus-green trail of smoke that hung in the air, like the exhaust of a missile shot from the constellation of Leo. This is where the meteor had originated.

Robert Frost wrote of the Leonids:

Did you sit up last night (the Magi did)
To see the star shower known as Leonid
That once a year by hand or apparatus
Is so mysteriously pelted at us?
It is but fiery puffs of dust and pebbles,
No doubt directed at our heads as rebels
In having taken artificial light
Against the ancient sovereignty of night.

As a backyard astronomer himself, Frost knew the stars. He got his first telescope at age 15, from an unlikely source — by selling magazine subscriptions. *The Youth's Companion* was a popular magazine over its long and varied history, beginning in 1827 as a small publication aimed at the religious market, to its demise in 1929 as a magazine for children and families. In its pages were stories of adventure, poems, and tales of science, meant to inspire young minds and authored by some of the best writers of the nineteenth century. Emily Dickinson and O. Henry both found homes here, and issues of the magazine were eagerly anticipated and read by the entire Frost household. Once a year, the magazine ran a promotion for readers who signed up the most new subscribers, offering an array of enticements from microscopes to air rifles.

In 1889, the enticement was a new telescope.

Robert Frost was a teenager when that first instrument arrived. A small French refractor not much larger than a pirate captain's spyglass, it sat in an extra bedroom, where Frost had draped a black curtain over a window, now open for stargazing. The telescope inched out through a hole cut in the drapery that formed a kind of home observatory, and with it began a poet's inspiration.

Like most amateurs, he began to read, teaching himself the constellations from books that he found at the Lawrence Public Library. He brought home star charts and atlases, and he tested his skills against the sky, as every observer does eventually, trying to match paper objects to the real thing.

He was also writing poems.

Writers are made, not born. The spark that begins a career is often like the match that ignites a forest fire, small and unnoticeable before the blaze takes over. For Robert Frost, the fire was nature unveiled, and soon astronomy began its silent claim on his writing. In December of 1891, he penned an editorial for his Lawrence High School newspaper. It read in part:

> What the school wants now is a telescope — mighty and far-reaching. . . .
>
> . . . The routine of school life fed entirely on books is unspeakably monotonous, so monotonous, in fact, that we become so depressed as to be uncertain one day whether or not we know what we learned yesterday; a little real observation would stand out of all this blackness as the moon seems to stand out of the darkness when looked at through a telescope; it makes the darkness seem pleasant.

The night sky would persist as an image for Robert Frost throughout his entire life. He would go on to write classic poems such as "A Road Taken" and "The Star-Splitter," but no writing is perhaps more beloved by astronomers than that of a poet in his own backyard, under the spell of night, as meteors arc overhead.

A fusillade of blanks and empty flashes
It never reaches the earth except as
   ashes.

These are the Leonids. They arrive every year in November, as sure as the locust, as dependable as the San Juan Capistrano swallows. Named after the radiant point in the constellation Leo, where the storm appears to originate, these meteors are in fact the rubble of an orbital comet, Tempel-Tuttle, discovered in December of 1865. Meteors are, as Frost had speculated, pieces of that comet — chunks of stone or iron that can range from the size of a baby pea to that of a golf ball. They come hurtling into the earth's atmosphere like a stunt car without brakes.

A few vaporize with great fanfare, sometimes leaving luminous ion trails, sometimes even breaking up; but what is most amazing about the Leonids is their sheer numbers. Swarms of meteors can overwhelm an observer, as they overwhelmed many in 2001 during the great Leonid storm. These meteors came relentlessly, a fierce rain of light that made some think that they were witnessing the end of the world.

In 1833, the Leonids terrorized the Eastern Seaboard. Illustrations from the period show meteors all over the sky, like snow, the exact metaphor conjured up by the writer Agnes Clerke in her description of the event:

On the night of November 12–13, 1833, a tempest of falling stars broke over the earth . . . the sky was scored in every direction with shining tracks and illuminated with majestic fireballs. At Boston, the frequency of meteors was estimated to be about half that of flakes in an average snowstorm.

According to the Natchez *Courier*, "The sky was so light that upon first awakening many thought that the city was on fire." People prayed for deliverance.

A few hours later, it was all over.

Few observers of my generation had ever seen anything resembling a meteor storm. As a boy, I had read about the 1833 storm and examined the woodcuts, but I was always disappointed by the reality of my own skies. A few meteors, darting under light-drenched heavens, would rarely get my attention before I went back to sleep. The Leonids peak quite early, around 3 a.m., and I haven't always had the dedication to climb out from under warm covers to see them.

I missed the great storm of 2001, sadly, but I wouldn't miss a second. Was this an encore? In all the newspaper and magazine coverage, the word was encouraging. The Leonids would surpass all others, some wrote, even those of 2001. A dicey call, to be sure; predicting meteor storms is a lot like figuring out who will win the Super Bowl five years from now — a real stretch. But there were new computer models, dust-trail analysis that promised to correlate the amount of material still floating in orbit against the path of the earth. Scientists claimed that we were going head-on in a stream of the thickest material yet, a stream still fresh from 1833 itself.

There was a glitch. The moon. A few hours from full, the moon would be low to the west, yet still bright enough to wash out an entire sky. So what? The Leonids are bright.

Not that bright.

I watched as a few tumbled overhead. The faint ones were missed by these eyes, so I can't comment on whether we went through a good stream or merely a nonexistent one, but I wasn't disappointed. The night sky had pulled me out of bed again, and that was enough. There are meteors to be seen on every clear night, I knew, as many as two or three an hour in dark skies. There are also other meteor showers: the Perseids in August, which is always strong; the Geminids in December; the Aquarids in May; and others. This is the true wonder of the night sky. It's forgiving.

★ ★ ★

November brings another gift. Later that evening, only hours after the Leonids, there was a total eclipse. This too is the hallmark of the heavens. One celestial event begets a second, a double feature of splendor.

I love lunar eclipses. I recall my first, in the late '60s and I have observed as many as I could, including my last one, a great dark eclipse in 1975. Eclipses are rated by their color, often copper or light brown, though sometimes autumn red, and by the darkness of the event. The shadow of the earth crossing the moon during an eclipse looks like the opening credits of a bad horror film with werewolves on the prowl, zombies rising up from old graves. No doubt the ancients felt the exact same way. Eclipses demand respect. They also inspire fear.

I heard this fear growing up, in my mother's voice. She would consult her eclipse tables, tabulated by a professional astronomer, no doubt, for these calculations were accurate, and she would shake her head.

"Not good. An eclipse in Aries, opposite of Mars."

Something would be shaken loose. A skirmish. A small catastrophe.

I tried to be reasonable with her, scientific.

"It happens often, Mom. You should watch it."

But my mother refused. At least in front of me. Then I caught a glimpse of her upstairs, tugging at the blinds. She was curious, of course, as much about the event perhaps as about the astrological outcome, which never came to pass.

Lunar eclipses are natural events, even though they may appear unnatural. Periodically, often in the spring and autumn, the earth will find itself in the same lane as the moon. Our planet becomes a speeding maniac that blocks out the light of everything close by, including the moon's. What we see in a lunar eclipse is the earth, several times larger than the moon, and hogging up all the road by making a path across the desolate lunar mug. It's lonely without the sun's light, and the moon probably feels it, along with the cold of its orbital big brother.

Typically, lunar eclipses don't last long — hardly more than a few hours. But the shadow that crosses the moon varies. The darkest part, the umbra, is caused by the deepest portion of the earth's ball, though there are eclipses when the umbra misses the moon completely. These are penumbral eclipses, and I'm looking at one now.

But there are also clouds tonight, thin layers of haze that work in front of the moon and then, as a tease, back off. This goes on throughout the eclipse as I watch patiently. Forget about photography; I am simply content to let the hours pass.

They do. For two consecutive nights I have been up late. The circadian rhythms of night and day are reversed for me. I stir at twilight, like Bela Lugosi from his coffin, only to reel at dawn, though I'm not alone. In Manhattan, more than one hundred thousand people — more folks than live in some cities — work during the late evening, when the rest of the East Coast is sleeping. For generations, the graveyard shift has been a time-honored production tool, cranking out widgets as the world dreams.

America, I sense, is staying up later, getting less sleep. It may have started with the first generation to have gone to college in such large numbers. University life has made night owls out of many of us with its studying, all-night libraries, and parties; sleep just didn't seem to be necessary. It continued with late-night television, cable access, the Internet. Young children will nip this in the bud pretty fast, though I notice that the effect could be hereditary. At night, I hear my oldest daughter in her room, singing or playing with her dolls. She's a night owl too, sometimes opening her window and asking me what I'm doing outside.

"Is it clear?"

"Yes. A beautiful night."

I want to invite her out, in her jammies, and show her the majesty of the sky, but I don't. It's a school night. Parents have to toe the line, keep the rules enforced, or else the house becomes a decaying orbit, manacled to gravity.

"What are you looking at?"

"The stars. Now off to bed."

She listens and goes to sleep. But someday she will be outside, in the cool of night, sitting on her porch as her own children rustle and ask for water. The circle will close the moment she looks skyward, as all of us must do eventually, and wonder: What am I looking at?

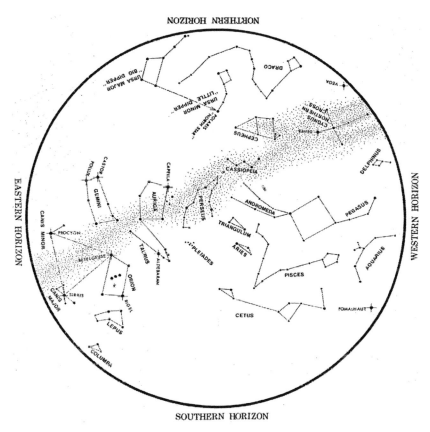

EASTERN HORIZON

WESTERN HORIZON

SOUTHERN HORIZON

THE NIGHT SKY IN DECEMBER

December

The stars will form pictures, tell stories, if you let them. They will whisper ancient tales from another night, centuries ago, an evening of kings and queens, of royal hunts and heroes that are now immortalized in the skies, an ancestral link to every past on Earth. All our forebears gazed at the same stars, a constant that has eluded us in our lives on this planet. Cities change, but the stars above them do not, and when we look up, we are looking through the lens of passing time. A distant relative of mine might have noticed the tug of Orion on another December night, in Italy, two hundred years ago, and the thoughts that turned in his mind were perhaps no different from my own.

He might wonder, as every schoolchild does, about the evening and why it is so long in winter, almost desperately so. The solstice. It arrives every December, shortly before Christmas, a day that seems to end before it even has a chance to begin. The astronomical reasons for this are simple, as the solstice represents the point where the north pole is at its greatest tilt away from the sun. But for the backyard astronomer, the winter solstice is Christmas early — a gift of the canopy of night given so everyone can appreciate it before going to bed.

As the sky darkens around dinnertime, stars appear one by one. Capella, now almost overhead, is the brightest, and it signals that the waters of the autumn sky are receding, as all floods must do. The river Eridanus, a string of stars that form an oxbow lake near Orion, has rushed in from the southeast, but it is no longer a river. Its banks are steep, its waters shallow. Orion can now safely cross, stepping on rocks and wading through trickling rapids. He moves fast. He has to. There is a bull in the forest, a mad stray from the herd.

Like many hunters, Orion stalks his prey with great respect. There is a relationship between hunter and hunted, different sides of the same coin, that few of the other constellations can understand. But Orion can. He is linked, year after year, cycle after seasonal cycle, with this animal. Each winter, the Bull squares off in exactly the same stance, as does Orion, two victims of the movie *Groundhog Day*, the replaying of a battle that is at once new and terribly ancient, waged until eternity, or until the earth's procession stops.

And it will. In time, Orion will no longer be the hunter that we recognize and honor but a scattering of stars, open to the mythmakers among our distant progeny. The cycles of the heavens, seemingly infinite, are temporary. They will change too, as we ourselves will, with seasons that pass in the same manner, only longer. This is the sheer vastness of change in the universe, the distance between what was and what will be, not in the staccato-voiced words of Rod Serling on television, but in the slow unfolding of time. Change is everywhere, though we notice it last within ourselves.

For the past year, I have found myself in both familiar and strange territory. Though the heavens have remained the same since I was boy, I have not. The meandering of individual lives is a curious one and not easily explained, as we veer from our origins and set off, as a ship does, across a trackless sea. Stars were back in my life again, and there was now also a place from which to observe them.

The observatory roof was finally up and working. When I pulled on the sliders, the two retractable halves cleanly separated as the panels of a magician's box would, revealing a telescope inside. When my wife first saw this trick, she grinned, as much from relief as from surprise.

"It actually works. Can you believe it?"

I shrugged, knowing I would feel better after the first rain. It soon came.

The rain was late, a cold, spitting precipitation that stung faces, clogged dirty gutters, and leaked on telescopes. For the first time I sat inside with my roof closed, the observatory already painted and finished off, or so I hoped. The water began to hit the roof, like blasting rounds of firecrackers. It was loud in there, but it was also dry. The roof was holding all the water at bay, and I laughed to myself, more shocked than confident, that I wasn't somehow under Niagara Falls.

Then it began.

First, a trickle. The water began to spread out between the two halves, then it dribbled down. I tugged on the two roofs, pulling them closer as one would the lapels of a jacket. It didn't help. Soon there were five or six leaks.

Water will find any crack and go there. It will flow until it doesn't, weaving through rock, sand, and dirt, or around plastic, to rendezvous with the lowest point. Water and gravity are linked, two combatants on opposite ends of a battle to shape creation. Roofs have to be pliable, crafty. A good carpenter will anticipate the flow of rain and plan accordingly, with gutters, rain breaks, a steep pitch. I had none of the above.

My roof was pitched at 4/12, which meant that it rose four inches for every foot — about the minimum for building code. But it wasn't a steep roof by a long shot, nor did I have the virtue of gutters to catch the rain and hurl it away from the fascia. In the rain I was exposed. The roof took everything nature had to offer and more as the unexpected quirks in my construction were tested. The separate halves of my roof sent trickles weaving left and right like children on a waterslide.

"It leaks," I told my brother-in-law.

"You cut the roof in two. You were expecting — ?"

"Well, I wasn't expecting a shower."

"Did you caulk?"

Sure, halfheartedly.

"Then do it again."

I did as he told me. An entire tube went into the roof, and I found a few more small gaps. This helped with the next rain — a real pounding — and I used the opportunity to mark all the places I'd missed. After another rain I was set, and the roof hardly leaked anymore, except in one place, a troublesome area by the door. I tried weather stripping around the jamb, stuffing some chips of wood inside the cracks. Everything. Then it hit me. It wasn't the door at all.

The rain, I discovered, was being caught by the bottom of the garage track that hung on the cantilevers outside the observatory, and it ran under the roof, causing a slow drip. How was that possible?

"Are the tracks level?" asked Jim.

I had no idea. The top of the frame was level, of course, but I never measured the sliders, assuming that they would mirror the top of the frame. They didn't. I was off by an inch.

My old inch that I lost on the bottom of the foundation had finally caught up with me, on top. The mistakes of the past never completely leave us; we merely become friends with them. Such was my inch. The roof still opened. But that would soon change.

A week later, snow came. It was a typical New England snow — heavy and wet. The observatory was now a giant heap of white, only higher off the ground, like the scoops of an ice cream sundae. Soon there was wind, songs that I never noticed before, new calls in the architecture of night. Miraculously, the wind had poked holes in the clouds. A thicket had broken, and now a front was barreling in, clear sky. I dug out the door of the observatory and walked in, reaching for the roof.

Nothing happened.

The roof was frozen shut. Ice apparently likes plastic, and no amount of pulling could move it.

"You're kidding me," said my wife.

"It won't budge."

"I would think it'd open right up."

One roof, sure. But two?

"It's cold, anyway. Stay inside, watch a movie."

But I already had a hair dryer in my hand.

The secret is this. The stars are there only if you notice.

The first night of using my completed observatory, roof open, telescope catching ancient photons, was exactly how I had always envisioned it, all those many years ago as a child. I walked in, after knocking off ice and melting the rest, then pulled, and the stars revealed themselves — a planetarium in my own yard.

And what stars. High overhead is the constellation of Andromeda, still perched on her rock. Neither Aquarius nor the fish in the sea can save her. For that she will need a hero, a boy that has slain Medusa, the snake-haired woman who turned people to stone with only a passing glance. But Perseus was smart. He used Medusa's magical strength against her, turning his shield into a mirror, and now the woman's head is a trophy, an emblem of his own courage and cunning.

Another threat looms. Andromeda wails from a distant island. The shoals are dangerous, thick with sharks and killer whales. Cetus, the lord of the deep, circles the island, intent on a human meal. There is no boat for Perseus that could withstand the voyage, nor can he swim. His only hope is Pegasus, the winged horse, whose rule of the skies equals that of Cetus, but with one exception. Pegasus is a hero in waiting.

Finding Perseus is easy. Simply follow the square of Pegasus up the eastern wing to Almach, or Gamma Andromedae, the last star in the constellation. Almach is one of the finest double stars in the heavens, two beacons of yellow and blue that shine brightly in the eyepiece, pointing the way to Mirfak, or Alpha Perseus. From there a stargazer can locate a set of objects. One is a star cluster, or actually two clusters, and these can easily be seen with the naked eye. This is the famous double cluster, NGC 869 and 848, which is located between Mirfak and the smallest part of the **W** in Cassiopeia.

NGC stands for New General Catalog, an inventory of non-stellar objects first assembled by Johann Dreyer in the late nineteenth century. The pages are replete with wonderful sights, clusters and galaxies, like my favorite, NGC 2903 in Leo, and also a few bright objects. But none is any better than the double cluster. Through binoculars or a small telescope, the stars condense into two balls, two diamond settings trying desperately to outdo each other. The double cluster is one of those objects an amateur never tires of looking at. Like the Orion Nebula, the Hercules cluster, and the galaxy of Andromeda, the double cluster is high on the observational hit list. I saw it for years as a boy, and now the view continues to amaze me. Through an eyepiece, even with a small telescope, I can see color — these are blues and smart yellows that seem to make the stars sparkle.

A wide-field refractor with a low-powered eyepiece offers perhaps the best view. Both clusters can be nicely framed within the glass and compared, a beauty pageant in the sky. Which is more lovely? NGC 869, the western clump, which owns a few more stars; or NGC 848, the more condensed twin? Only a pair of binoculars and your own attitude can decide this.

Traveling south, an alert stargazer should be able to pick up the recognizable outline of Perseus, a constellation that resembles the loose fit of a cape. If we were naming the constellations today, Perseus might be a hooded bank robber or a goblin, with his head and spindly arms. Certainly he looks scary, but not as scary as the name of the second brightest star in Perseus — Algol. This is the Demon Star.

There is a sinister feel here, and it only takes a few days to notice it. Algol will wink from a bright 2.1 magnitude down to a much dimmer, though still easily visible, 3.4 magnitude every 2.86 days, making a casual observer wonder: Is there something going on with this star?

Yes, there is.

Algol is an eclipsing binary star, two stars actually, that was first noticed, officially, in 1667. But keen-eyed Chinese and Arabs, great astronomers in their own right, probably observed it well before then.

Throughout the ages there have been dark associations with Algol, suggesting that its variability was known far before its discovery. In Hebrew, Algol was Rosh Ha Satan, or Satan's Head. For the Chinese it was Tseih She, the grimly translated Piled-up Corpses. Astrologers didn't feel any more comfortable with this star, calling it the most dangerous object in the sky.

So why all the fuss? Algol has in orbit a companion, a second star, that every three days blocks out a portion of Algol's light, like a dimmer switch, by passing into our line of sight. These are stars in eclipse, a constant game of cat and mouse with earthly observers, and we are far the luckier to watch it.

Finding Algol is easy. From the double cluster, travel south to Mirfak, the brightest star in Perseus, and continue for a thumb and finger length to Algol. Any farther south, and a stargazer crashes into the Pleiades, or Seven Sisters. This too is one of the most recognizable objects in the sky to non-astronomers, and many children have tested their eyesight by picking out the seven stars, an optical test for the ancients as well.

With the naked eye, the Pleiades resemble a small cocoon just north of Aldebaran and the **V** of Taurus. But in binoculars, the Pleiades come to life. These are young stars, newly born infants still thrashing in the cradle. Circling the hundred or so stars that constitute the Pleiades system are faint wisps of nebula, ionized gas that has contracted into the atomic balls that now make up the big seven, including Alcyone, the brightest star, whose light anchors the cosmic riff on another constellation, one that is clearly copyrighted.

The Big Dipper.

Through binoculars, the bright stars of the Pleiades resemble a second, short-handled dipper, I've always thought, with a bowl shaped exactly like the famous original, only in miniature. Alcyone mans the top of the bowl while Merope, a star with the most surrounding nebulosity, anchors the southeastern bowl. The little bowl climbs from the winter sky on cold evenings, inspiring imaginations.

The rising and setting of the Pleiades have long been associated with a variety of traditions, from the opening of the ancient Greek naval season to the revelry of Allhallows Eve in the Middle Ages. They also seem to have inspired fear and awe, but today the Pleiades are harbingers of only one thing. Winter. And yet, when I look at them this cold evening, all I can see are stars — and something else, in Taurus, high up, a memory of long ago.

The memory comes from Saturn, glowing above. What is strange is that Saturn is now in the same constellation, Taurus, as when I first became serious about astronomy — twenty-nine and a half years ago. This is one entire orbit around the sun, but half a human lifetime for many. The thin ticking of the clock these years has been for me, it seems, a point that isn't lost in translation. I am older by thirty years. But Saturn goes about his merry way.

In a telescope, Saturn rarely disappoints. The rings are now open as wide as possible. This happens every thirty years or so, from pole to pole, as the tilt of the Earth-Saturn environment changes, giving the observer the best view of the ringed planet available. A racetrack with lanes, the rings are separated into striations, easily seen with most backyard instruments. The widest and darkest of these, Cassini's Division, was named after Jean-Dominique Cassini, an Italian who discovered it in 1675. Though obvious in a backyard telescope, Cassini's uncovering of the ring structure was no small feat.

From the first observation by Galileo in 1610, controversy surrounded his findings. Saturn, he observed, wasn't one planet but three, with tiny, attending globes on each side. And these attendants never left the planet's side. Two years passed. Galileo pointed his small refractor skyward again and was shocked. The two extra planets were gone.

Every fifteen years, Saturn loses its rings for a short period. They tilt into our line of sight and appear to evaporate, though in larger telescopes a thin line remains. Galileo couldn't see this. He could only observe one ball now, instead of three. What had happened?

"Have they vanished or suddenly fled?" he wrote. "Has Saturn perhaps devoured his own children? Or were the appearances indeed illusion or fraud, with which the glasses have so long deceived me, as well as many others to whom I have shown them?"

Bingo.

Galileo never knew it, but today we do. The two attendants were simply out-of-focus rings, symptoms of bad spherical and chromatic aberrations due to poor optics.

He continued, "The shortness of time, the unexpected nature of the event, the weakness of my understanding, and the

fear of being mistaken, have greatly con-
founded me."

This is the dark night of the soul for an
astronomer. A scientist under pressure
because of the obvious silliness of his observations will stick by them if the methods are
sound. But it takes others to prove or dis-
prove a theory. For Galileo it was Christiaan
Huygens who, after discovering Titan, one
of the myriad of Saturnian moons, offered
up his explanation. Saturn had a symmet-
rical ring circling it.

A ring? Was this a worse insanity? Who
had ever observed a ring around anything
before? And how would it possibly be attached?

Continued observations proved Huygens
to be correct. And now the ring system of
Saturn, so beautiful as to almost take the
breath away from my children, tantalizes all
those who look up.

My children see the bands of Saturn,
usually a light one running around the
equator and a darker one, on top at the
pole, and they smile. This is the seduction
of Saturn. In thirty years they will watch as
Saturn graces Taurus once again, linking
past and present, memory with anticipation, except this time the clock will tick for
them.

"It looks striped," quips my youngest, "like a beach ball."

"That's only the belts. They're made of gas."

"Gasoline? That's silly."

I correct her. Hydrogen, ammonia, methane. The belts are cold and probably just as deadly.

"But I see little stars."

"Moons."

Saturn has more moons than I can possibly name or number. As a boy in the late '60s, I knew every moon by name, but so many have now been discovered that I simply can't keep track. I don't think anyone can. Moons are multiplying in our solar system, like mice in a maternity ward.

"You can see four or five pretty easily."

She counts. Titan, the largest; Rhea; Dione; Tethys; and maybe Iapetus.

Iapetus has an interesting problem. From western to eastern elongation, or the farthest stretches on either side of a planetary orbit, Iapetus can vary in brightness by as much as two magnitudes. The reasons are complex and only partially understood. Iapetus may be like Jupiter's Io, subject to volcanism, or it may have large spreads of light and dark deposits, probably ice, that alter the reflective rays of the sun. Either way, Iapetus fades in and out of my five-inch, depending upon the day of the week.

My daughter looks up. She's hooked.

"You should take a picture of this, Daddy."

A picture.

For the past year, I have tinkered with photography, taking pictures of a few bright galaxies, star clusters, and the moon. But now, with Orion rising again over the trees, I hear the old calling, one that pushed me on this journey in the first place, to photograph the Horsehead Nebula.

I set up my camera and all the attending equipment — an off-axis guider, illuminated eyepiece, and motorized focuser — and I stuff a fresh set of batteries into the clock drive. This is the moment of opportunity.

Two rolls go by on three consecutive evenings. When I get them back from the developer, I find the same old problems with guiding, trailed images, or blurry slides. It's difficult to focus a camera on a telescopic object, much more difficult than you would imagine. The light is faint, and staring through a camera's viewfinder doesn't help any, unless it's a bright star. What comes back to the eye is often blackness. So a photographer is left with a series of tricks to focus the camera, using specialized camera screens, telescope masks, and even computer software to help sharpen the view through a telescope.

Some pictures that I got back are simply dreadful. There is the bright star Zeta Orionis and little else, or just a blob. To get a good picture, I decide, I need to closely guide the image, steer it by hand. I also need a break.

On my last roll, I get lucky. One frame is covered in light and something else. A faint imprint. It's small, like a dark seed, but unmistakable. It's the Horsehead Nebula. I have finally bagged this elusive object, not with the mighty glass of the Keck, or with the seasoned cameras of astrophotographers far more talented than I, but with my own two hands.

Light, in a year, will traverse almost six trillion miles, or a quarter of the way to Proxima Centauri, our closest stellar neighbor. Its speed is measurable, finite. But what is the velocity of memory, of reclaiming our past? A year earlier I was rushing in from evening chores, hardly bothering to look up and take notice of the stars above me. Exiled from the love of my childhood, I simply did what was natural. I pretended the stars didn't matter. No longer. The peace that comes from observing the heavens is something hardly quantifiable, though mystics offer a few names — prayer, *zazen*. I just call it an evening under the stars.

The sky is always changing, forcing us along through its seasons of night. It never fails us, this turbulent cosmos, and this perhaps is its final strength. The universe comes to us, ancient and shopworn, with its own meaning attached. Or we are free to add our own. My mother did. For her, the heavens were portents of some potential future, a fork marked between the road of what was possible and what was likely, a view I never bought. But now I wonder: Was she right? The stars have influenced me, it turns out, as powerfully as they influenced my mother, not by an invisible plea to control my life, mold a silent destiny as a parent does, but by their sheer presence.

Families offer presence too, of course, the solid grounding of a father and mother, and when one of them dies, the vacuum left is a hole that few of us can anticipate. No influence in the universe, astrological or scientific, is stronger than the bond between a parent and a child. It transcends the gravitational pull of neutron stars, rejects the hydrogen shells of rushing nebulae, laughs at black holes. The influence is a simple one: love.

Love can bring us a long distance, if we take notice. I finally did. Throughout a celestial season, I had spent twelve months noticing, watching comets and asteroids, faint galaxies, sunspots, and distant stars. Creation unfolding. Some may witness this unfolding of the universe, like a gathering of planets, as comforting, a creation that cares enough to influence us, and they interpret it as such — a mystery with human fingerprints. Others scoff and find no connection where none was intended. Both sides miss the point, I think. And therein lies perhaps the greatest mystery: not how strange it is for the universe to unfold, but rather, that there is a universe to unfold at all.

Selected Bibliography

Abrahams, Peter. "Henry Fitz: American Telescope Maker." *Journal of the Antique Telescope Society,* vol. 6 (Summer 1994).

Ackerman, Diane. *A Natural History of the Senses.* New York: Vintage, 1991.

Alberg, Henry, ed. *Maria Mitchell: A Life in Journals and Letters.* Clinton Corners, NY: College Avenue Press, 2001.

Allen, Richard Hinckley. *Star Names.* New York: Dover, 1963.

André, C., G. Rayet and A. Angot. *L'astronomie pratique et les observatoires en Europe et en Amérique.* Paris: Gauthier-Villars, 1877.

Aveni, Anthony F. "Archaeoastronomy in the Maya Region." *Archaeoastronomy,* no. 3 (1981).

Aydin, Sayili. *The Observatory in Islam.* New York: Arno Press, 1981.

Baker, Daniel W. *History of the Harvard Observatory During 1840–1890.* Cambridge, Mass.: Harvard University Press, 1890.

Baker, Jean H. *Mary Todd Lincoln.* New York: W. W. Norton, 1987.

Bauval, Robert, and Adrian Gilbert. *The Orion Mystery.* New York: Crown, 1994.

Bell, Louis. *The Telescope.* New York: McGraw-Hill, 1922.

Bell, Trudy. "In the Shadow of Giants: Forgotten Nineteenth Century Telescope Makers." *Griffith Observer,* September 1986.

Belting, Natalia. *The Moon Is a Crystal Ball.* Indianapolis: Bobbs-Merrill, 1952.

Bohm, David. *Wholeness and the Implicate Order.* London: Ark, 1983.

Brecher, Kenneth, and Michael Feirtag, eds. *Astronomy of the Ancients.* Cambridge, Mass.: MIT Press, 1979.

Burnham, Robert, Jr. *Burnham's Celestial Handbook,* 3 vols. New York: Dover, 1978.

Buttman, Gunther. *The Shadow of the Telescope: A Biography of John Herschel.* New York: Charles Scribner's Sons, 1970.

Byron, Lord (George Gordon). *Selected Poems.* New York: Gramercy Books, 1994.

Chapman, Allan. *The Victorian Amateur Astronomer.* Boston: Wiley, 1998.

Clerke, Agnes. *Popular History of As-*

tronomy During the 19th Century. London: Adam and Charles Black, 1908.

Cline, Ann. *A Hut of One's Own.* Cambridge, Mass.: MIT Press, 1997.

Cook, Raymond. "Robert Frost: Poetic Astronomer." *Emory University Quarterly,* vol. 16, no. 1 (Spring 1960).

Davis, Herman S. "David Rittenhouse." *Popular Astronomy,* vol. 4, no. 1 (July 1896).

Dedication of Palomar Observatory and the Hale Telescope. California Institute of Technology, June 3, 1948.

——. "John Quincy Adams, The Smithsonian Bequest and the Founding of the U.S. Naval Observatory." *Journal History of Astronomy,* vol. 22 (February 1991).

Dick, Steven J. "Sears Cook Walker and the Philadelphia High School Observatory." *Bulletin of the American Astronomical Society,* vol. 22 (1990).

Donald, David Herbert. *Lincoln.* New York: Simon & Schuster, 1996.

Donnelly, Marion Card. *A Short History of Observatories.* Portland, Ore.: University of Oregon, 1973.

Emerson, Ralph Waldo. *Emerson in His Journals.* Ed. Joel Porte. Cambridge, Mass.: Harvard University Press, 1982.

Faits, Ed. "Amasa Holcomb: Pioneer Telescope Maker." *The Reflector,* June 1991.

Ferguson, Kitty. *Tycho and Kepler.* New York: Walker, 2002.

Ferris, Timothy. *Seeing in the Dark.* New York: Simon & Schuster, 2002.

Florence, Ronald. *The Perfect Machine.* New York: Harper, 1995.

Franch, John. "Charles Tyson Yerkes." *University of Chicago Alumni Magazine,* February 1997.

Frost, Robert. *Letters to Louis Untermeyer.* Ed. Louis Untermeyer. New York: Holt, 1963.

———. *The Poetry of Robert Frost.* Ed. Edward Connery Lathem. New York: Owl Books, 1979.

Galilei, Galileo. *Discoveries and Opinions of Galileo.* Translated with an introduction and notes by Stillman Drake. New York: Anchor, 1957.

Goderya, Shaukat, Suhail Farooqui and Ali-Mohammad. "Feasibility Report for the Establishment of an Astronomical Observatory." United Nations Development Program, 1998.

Goff, John. *Robert Todd Lincoln.* Oklahoma: University of Oklahoma, 1968.

Golub, Leon, and Jay M. Pasachoff. *Nearest Star: The Surprising Science of*

Our Sun. Cambridge, Mass.: Harvard University Press, 2001.

Greenslet, Ferris. *The Lowells and Their Seven Worlds.* Cambridge: Houghton Mifflin, 1946.

Hawkins, Gerald. *Stonehenge Decoded.* New York: Doubleday, 1965.

Heilbron, J. L. *The Sun in the Cathedral.* Cambridge, Mass.: Harvard University Press, 1999.

Hindle, Brooke. *The Pursuit of Science in Revolutionary America, 1735–1789.* Chapel Hill, N.C.: University of North Carolina Press, 1956.

History of AAVSO. AAVSO Web site.

History of Astronomy at UNC. University of North Carolina Web site.

History of the Cincinnati Observatory. Cincinnati Observatory Web site.

Hoffleit, Dorrit. *Astronomy at Yale, 1701–1968.* New Haven: Yale University, Connecticut Academy of Art and Sciences, 1992.

Hoskin, Michael, ed. *Cambridge Illustrated History of Astronomy.* Cambridge, U.K.: Cambridge University Press, 1997.

Jerome, John. *Stone Work.* New York: Viking, 1989.

Kelleher, Florence M. "George Ellery Hale: The California Years." Yerkes Observatory Virtual Museum. astro.uchicago. edu/yerkes/virtualmuseum

———. "George Ellery Hale: The Early Years." Yerkes Observatory Virtual Museum. astro.uchicago.edu/yerkes/virtual museum

King, Henry C. *The History of the Telescope.* New York: Dover, 1957.

Kronk, Gary. http://cometography.com

Krupp, E. C. *Echoes of the Ancient Skies.* New York: New American Library, 1984.

———. "The Observatory at Guo Shou Jing." *Griffith Observer*, August 1982.

Lankford, John, ed. "Astronomy's Enduring Resource." *Sky & Telescope*, November 1988.

———. *History of Astronomy.* New York and London: Garland Publishing, 1997.

———. and Rickey L. Slavings. "Gender and Science: Women in American Astronomy, 1859–1940." *Physics Today*, March 1990.

———. and Rickey L. Slavings. "The Industrialization of American Astronomy, 1880–1940." *Physics Today*, vol. 49, no. 1 (January 1996).

Levy, David H. *Guide to Observing and Discovering Comets.* Cambridge, U.K.: Cambridge University Press, 2003.

———. "Miracle at Birr Castle." *Sky & Telescope*, January 2004.

———. "Walden of the Sky." *Sky & Telescope*, September, 1995.

Levy, Matthys, and Mario Salvadori. *Why Buildings Fall Down.* New York: W. W. Norton, 1992.

Ley, Willy. *Watchers of the Sky.* New York: Viking, 1966.

Loomis, Elias. "Astronomical Observatories in the U.S." *Harper's New Monthly Magazine*, vol. 13, June 1856.

———. *Recent Progress of Astronomy, Especially in the United States.* New York: Harper Brothers, 1850.

Love, James L. "The First College Observatory in the United States." *Sidereal Messenger*, vol. 7, no. 10 (December 1888).

The McDonald Telescope: Commemorating the Dedication and Formal Opening of the McDonald Observatory at the University of Texas. Caxton, for Warner and Swasey Company, May 5th, 1939, Cleveland, OH.

Mackenzie, Dana. *The Big Splat.* Boston: Wiley, 2003.

Marshall, Oscar Seth. *Journeyman Machinist En Route to the Stars.* Taunton, Mass.: William Sullwold Publishing, 1979.

Maslikov, Sergey. "Amateur Astronomy in Russia." *Sky & Telescope*, September 2001.

McLaughlin, Jack. *Jefferson and Monticello.* New York: Holt, 1990.

Milham, Willis I. *Early American Observatories.* Williamstown, Mass: Williams College, 1924.

——. *The History of Astronomy at Williams College and the Founding of the Hopkins Observatory.* Williamstown, Mass.: Williams College, 1937.

Misch, Tony, and Remington Stone. "The Building of Lick Observatory." University of California Observatories, 1998.

——. "James Lick: The Generous Miser." University of California Observatorics, 1998. mthamilton. uslick.org/public/history

Mitchell, Maria. *Journals, 1896.* eBook: Project Gutenberg, 2003.

Moore, Patrick. *Astronomical Telescopes and Observatories for Amateurs.* New York: Norton, 1973.

——. *The Planet Venus.* New York: Macmillan, 1957.

——. *Planet Venus.* London: Cassell, 2002.

Moore, Patrick, ed. *Practical Amateur Astronomy.* Guildford and London: Lutterworth Press, 1973.

Moser, Don. "A Salesman for the Heavens Wants to Rope You In." *Smithsonian,* April 1989.

Multhauf, Robert. "Holcomb, Fitz and Peate: Three 19th Century American Telescope Makers." *Bulletin 228: Contributions from the Museum of History and Technology.* Washington, D.C.: Smithsonian Institution, 1962.

Norse, J. E. "Observatories in the United States." *Harper's New Monthly Magazine,* vols. 47 (March 1874) and 49 (September 1874).

Northcott. "Early Accomplishments of Amateur Astronomy in Canada." *Journal of the Royal Astronomical Society of Canada,* vol. 59 (1966).

O'D. Alexander, A. F. *The Planet Saturn.* New York: Dover, 1980.

———. *The Planet Uranus.* London: Faber, 1957.

O'Meara, Stephen James. "Longfellow: Voice of the Night." *Sky & Telescope,* June 2001.

Osterbrock, Donald E., John R. Gustafson, and W. J. Shiloh Unruh. *Eyes on the Sky: Lick Observatory's First Century.* Berkeley: University of California Press, 1988.

Peek, B. M. *The Planet Jupiter.* London: Faber, 1958.

Peltier, Leslie. *Starlight Nights.* Boston: Sky Publishing, 1999.

Pendergrast, Mark. *Mirror, Mirror.* New York: Basic Books, 2003.

Petrunin, Yuri, and Edward Trigubov. "Dmitri Maksutov." *Sky & Telescope,* December 2001.

Pickering, David. "An Amateur Observatory." *Popular Astronomy,* vol. 38 (1936).

Poe, Edgar Allan. *The Science Fiction of Edgar Allan Poe.* Ed. Harold Beaver. London: Penguin, 1977.

Pollan, Michael. *A Place of My Own: The Education of an Amateur Builder.* New York: Dell, 1997.

Pomerance, Anita. "Hildene Restores Lincoln's Telescope." *Manchester Journal,* September 13, 2002.

Price, Fred W. *The Moon Observer's Handbook.* Cambridge, U.K.: Cambridge University Press, 1988.

——. *The Planet Observer's Handbook.* Cambridge, U.K.: Cambridge University Press, 2000.

Proctor, Mary. *The Romance of Comets.* New York: Harper Brothers, 1926.

Richardson, Robert D., Jr. *Emerson: The Mind on Fire.* Berkeley: University of California Press, 1995.

Ridpath, Ian, ed. *Norton's 2000 Star Atlas.* London: Longman, 1989.

Ronan, Colin A. *Galileo.* New York: Putnam, 1974.

Rosen, Edward. *The Naming of the Telescope.* New York: Henry Schuman, 1947.

Roth, Gunter D. *Handbook for Planet Observers.* London: Faber, 1966.

Rubincam, David Parry, and Milton Rubincam II, "David Rittenhouse: America's Foremost Early Astronomer." *Sky & Telescope,* May 1995.

Rufus, W. Carl. "Astronomical Observatories in the United States Prior to 1848." *Scientific Monthly,* vol. 19 (August 1924).

———. "Proposed Periods in the History of Astronomy in America." *Popular Astronomy,* vol. 29, nos. 7 and 8 (1921).

Safford, Truman Henry. "50th Anniversary of Williams College Observatory." Pamphlet. Williams College, 1888.

Sander, Werner. *The Planet Mercury.* New York: Macmillan, 1963.

Sheehan, William, and Stephen J. O'Meara. *Mars: The Lure of the Red Planet.* New York: Prometheus Books, 2001.

Sidgwick, J. B. *Observational Astronomy for Amateurs.* New Jersey: Enslow, 1982.

Sluiter, Engel. "The First Known Telescopes Carried to America, Asia, and the Arctic, 1614–39." *Journal of the History of Astronomy*, vol. 28 (1997).

———. "The Telescope Before Galileo." *Journal of the History of Astronomy*, vol. 28 (1997).

Sobel, Dava. *Longitude.* New York: Walker and Co., 1995.

Sperling, Norman. "When Comets Were Discovered from Newark." *Sky & Telescope*, August 1979.

Stebbins, Robert A. "Looking Downwards: Sociological Images of the Vocation and Avocation of Astronomy." *Journal of the Royal Astronomical Society of Canada*, vol. 75, no. 1 (1981).

Thom, A. *Megalithic Lunar Observatories.* London: Oxford, 2002.

Thompson, Lawrence. *Robert Frost.* 3 vols. New York: Holt, Rinehart, and Winston, 1970.

Tombaugh, Clyde. *The Planet Pluto: Out of Darkness.* Harrisburg, Penn.: Stackpole Books, 1980.

Toth, Jennifer. *The Mole People.* Chicago: Chicago Review Press, 1995.

Walker, Lester. *The Tiny Book of Tiny Houses.* Woodstock, NY: Overlook Press, 1993.

Walker, Merle. "How Good Is Your Observing Site?" *Sky & Telescope*, February 1986.

Walters, Raymond. Centenary, Cincinnati Observatory. *Science*, 1943.

Warner, Deborah Jean. "200 Years of Amateur Astronomy." Proceedings, ALPO convention, 1976. Las Cruces, NM.

Warner, Deborah Jean, and Robert B. Ariail. *Alvan Clark and Sons: Artists in Optics.* Virginia: Willmann-Bell, 1996.

Webb, T. W. *Celestial Objects for Common Telescopes.* 2 vols. Dover, 1962.

Wells, H. G. *The War of the Worlds.* New York: Bantam, 1988.

Willard, Berton C. *Russell W. Porter.* Freeport, Maine: Bond-Wheelwright, 1976.

Williams, Thomas R. "Astronomers as Amateurs." *Journal of the AAVSO*, no. 12 (1983).

———. "A Galaxy of Amateur Astronomers." *Sky & Telescope*, November 1988.

———. "Albert Ingalls and the ATM Movement," *Sky & Telescope*, February 1991.

Yeomans, Donald K. *Comets: History of Observation, Science, Myth and Folklore.* Boston: Wiley, 1991.

The Youth's Companion, July 4, 1889.

Zinszer, Harvey. "Famous Early American

Observatories." *Transactions of the Kansas Academy of Science*, vol. 47 (1944).

Further Reading

BOOKS AND STAR ATLASES

Amateur Astronomy by Patrick Moore (Norton)

The Backyard Astronomer's Guide by Terence Dickinson and Alan Dyer (Firefly Books)

The Binocular Stargazer by Leslie C. Peltier (Kalmbach)

Cambridge Encyclopedia of Amateur Astronomy by Michael E. Bakich (Cambridge University Press)

How to Use an Astronomical Telescope by James Muirden (Linden Press)

Norton's Star Atlas, edited by Ian Ridpath (Longman)

One Hundred Greatest Stars by James B. Kaler (Copernicus Books)

Peterson's Field Guide to the Stars and Planets by Jay M. Pasachoff (Houghton Mifflin)

365 Starry Nights by Chet Raymo (Fireside)

Touring the Universe with Binoculars by
  Philip S. Harrington (Wiley)

PERIODICALS

Astronomy
Kalmbach Publishing
21027 Crossroads Circle
P.O. Box 1612
Waukesha, WI 53187

Sky & Telescope
49 Bay Street
Cambridge, MA 02138

Mercury (bimonthly)
Astronomical Society of the Pacific
390 Ashton Avenue
San Francisco, CA 94112

Stardate (bimonthly)
Published by the McDonald Observatory
University of Texas
Austin, TX 78712

Griffith Observer
Griffith Observatory
2800 E. Observatory Road
Los Angeles, CA 90027

Journal of the British Astronomical Association
The British Astronomical Association
Burlington House
Piccadilly
London W1J 0DU

Journal of the Royal Astronomical Society of Canada
136 Dupont Street
Toronto, ON M5R 1V2

AMATEUR GROUPS

American Association of Variable Star Observers
www.aavso.org

Association of Lunar and Planetary Observers
www.lpl.arizona.edu/alpo

British Astronomical Association
www.britastro.org

The Astronomical League
www.astroleague.org

Astronomical Society of the Pacific
www.astrosociety.org

Royal Astronomical Society of Canada
www.rasc.ca

Acknowledgments

The author would like to thank the following: Bruce Bradley, Sara Carder, Tony Cook, Dorrit Hoffleit, Dorian Karchmar, Dr. E. C. Krupp, Barbara Lowenstein, Kim Monocchi, Peter Nelson, Dr. Jay M. Pasachoff, Dr. Vladimir Strilenski, Jeremy Tarcher, Jim, and my family, plus all the good folks at hardware stores and lumberyards everywhere for sharing their knowledge with me.

About the Author

Charles Laird Calia is the author of the novel *The Unspeakable*. He is a frequent contributor to *Sky & Telescope* magazine. A member of the American Association of Variable Star Observers, the Association of Lunar & Planetary Observers, and the British Astronomical Association, Calia lives in Connecticut with his wife and their two daughters.